T0137266

Multiscale Forecasting Models

Lida Mercedes Barba Maggi

Multiscale Forecasting Models

 Springer

Lida Mercedes Barba Maggi
Universidad Nacional de Chimborazo
Riobamba, Ecuador

Pontifícia Universidad Católica de Valparaíso
Valparaíso, Chile

ISBN 978-3-030-06950-6   ISBN 978-3-319-94992-5   (eBook)
https://doi.org/10.1007/978-3-319-94992-5

Printed on acid-free paper

This Springer imprint is published by the registered company Springer Nature Switzerland AG
The registered company address is: Gewerbestrasse 11, 6330 Cham, Switzerland

Specially to my parents Lidita and Guillito, my husband Oswaldo, my son Joe, and my daughter Shandy, who are my breath and my source of strength, my love, and my support.

Preface

This book contains conventional and competitive methods to forecast time series. The methods are presented in detail through relevant applications. Only discrete time series are used in order to approach the vast majority of time series applications.

It has been demonstrated that time series are valuable sources of information for supporting planning activities and designing strategies for decision-making. Multiple demands coming from diverse sectors such as, transport, fishery, economy, and finances are concerned with obtaining information in advance to improve their productivity and efficiency. Unfortunately, the presence of high variability in a time series is often observed, and this situation obeys to the presence of *trend, seasonal variation, cyclic fluctuation*, or *irregular fluctuation*.

During the last decades, numerous forecasting models to explain the behavior of an observed phenomena have been developed. Autoregressive integrated moving average (ARIMA) is oriented to forecast nonstationary time series. ARIMA implements D processes of differentiation for transforming a nonstationary time series into stationary time series; most time series are stationarized taking the first difference ($D = 1$). An ARMA(P, Q) model is the generalization of an ARIMA(P, D, Q) model, where P is the order of the autoregressive part (number of autoregressive terms) and Q is the order of the moving average part (number of lagged forecast errors).

On the other hand, nonlinear relationships are often modeled by nonlinear models. Artificial neural networks (ANNs) are popular techniques of artificial intelligence that are implemented in regression problems. ANN implementations imply to take some decisions, after several tests, about some parameters such as network topology, signal propagation method, activation function, weights updating, hidden levels, and numbers of nodes. A specific configuration cannot be generalized, even in similar studies.

Hybrid models are a recent solution to deal with nonstationary processes. Hybrid models combine preprocessing techniques with conventional linear and nonlinear models; two of them, widely implemented, are singular spectrum analysis (SSA) and discrete wavelet transform (DWT). These techniques extract components from observed signals, and the components are smoother than the original signal which improve forecast models.

Hankel singular value decomposition (HSVD) is a new strategy implemented to extract components of low and high frequency from a nonstationary time series. The proposed decomposition is used to improve the accuracy of linear and nonlinear autoregressive models. More precisely, the contributions of this work are the following:

- *Elemental content* about time series and conventional forecasting.
- *One-step ahead forecasting* based on Hankel singular value decomposition for *linear and nonlinear* forecasting models.
- *Multi-step ahead forecasting* based on Hankel singular value decomposition for *linear and nonlinear* forecasting models.
- *Multi-step ahead forecasting* based on multilevel singular value decomposition for *linear* forecasting models.
- *Singular spectrum analysis and stationary wavelet decomposition are* used to compare the performance of *HVD* and *MSVD*.

This contribution was validated with time series coming from traffic accidents and fishery stock. Six time series obtained from CONASET and SERNAPESCA were used; they are as follows:

1. Weekly sampling from 2003:1 to 2012:12 of *injured* persons in traffic accidents in *Valparaíso*.
2. Weekly sampling from 2000:1 to 2014:12 of *injured* persons in traffic accidents in *Santiago* due to *10 principal causes related to inappropriate behavior* of drivers, passengers, and pedestrians.
3. Weekly sampling from 2000:1 to 2014:12 of *injured* persons in traffic accidents in *Santiago* due to *10 secondary causes related to inappropriate behavior* of drivers, passengers, and pedestrians.
4. Weekly sampling from 2000:1 to 2014:12 of *injured* persons in traffic accidents in *Santiago* due to *causes related to road deficiencies, mechanical failures, and undetermined causes*.
5. Monthly sampling from 1958:1 to 2011:12 of *anchovy catches* at center-south coast of Chile.
6. Monthly sampling from 1949:1 to 2011:12 of *sardine catches* at center-south coast of Chile.

This work is structured as follows. Chapter 1 provides a literature review about time series, linear forecasting models, and ANNs. In Chap. 2, HSVD is presented. HSVD in conjunction with AR and an ANN are applied for one-step ahead forecasting of traffic accidents in Valparaíso. In Chap. 3 multi-step ahead forecasting via direct and MIMO strategies is presented. The performance of the proposed forecasters is compared with the decomposition technique widely used in SSA. Multilevel singular value decomposition is presented in Chap. 4. Forecasters are based on AR model and MIMO strategy. The performance of MSVD is compared with SWT; two experiments were performed for multi-step ahead forecasting of traffic accidents and fisheries stock.

This book assumes a knowledge only of basic calculus, matrix algebra, and elementary statistics. This book is aimed at practitioners and professionals, lecturers and tutors, academic and corporate libraries, undergraduates, and postgraduates.

Riobamba, Ecuador   Lida Mercedes Barba Maggi
April 2018

Acknowledgments

I am very thankful with Pontificia Universidad Católica de Valparaíso for the scholarship which definitely helped me to complete my postgraduate studies. My special thanks to Professor Nibaldo Rodríguez, member of the Computational Intelligence research group, for allowing me to benefit from his knowledge and experience.

My deepest gratitude to my dear Universidad Nacional de Chimborazo for giving me license and support to perform this study, which would not have been possible without their help.

Riobamba, Ecuador   Lida Mercedes Barba Maggi

Contents

List of Figures

List of Tables

Acronyms

ACF	Autocorrelation Function
ADF	Augmented Dickey-Fuller test
ANN	Artificial Neural Network
AR	Autoregressive
ARIMA	Autoregressive Integrated Moving Average
ARMA	Autoregressive Moving Average
BP	Backpropagation
CONASET	Chilean Commission of Traffic Safety
EIA	Energy Information Administration
GCV	General Cross Validation
HSVD	Singular Value Decomposition of Hankel
KPSS	Kwiatkowski, Phillips, Schmidt, and Shin test
LM	Levenberg-Marquardt
LS	Least Squares
MA	Moving Average
MAPE	Mean Absolute Percentage Error
mIA	Modified Index of Agreement
MIMO	Multiple input, multiple output
MLE	Maximum Likelihood Estimation
MLP	Multilayer Perceptron
mNSE	Modified Nash-Sutcliffe Efficiency
MRA	Multiresolution Analysis
MSE	Mean Square Error
NOAA	National Oceanic and Atmospheric Administration
Norm	Normalized
nRMSE	Normalized Root Mean Square Error
PACF	Partial Autocorrelation Function
PDF	Probability Density Function
RE	Relative Error
RMSE	Root Mean Square Error
RPROP	Resilient Backpropagation

SAD	Sum of Absolute Deviation
SAE	Sum of Absolute Error
SERNAPESCA	National Service of Fisheries and Aquaculture
SSA	Singular Spectrum Analysis
SVD	Singular Value Decomposition
SVM	Support Vector Machine
SVR	Support Vector Regression
SWT	Stationary Wavelet Transform
TSDL	Time Series Data Library
TWB	The World Bank Open Data
WN	White Noise

Chapter 1
Times Series Analysis

1.1 Time Series

Data continuously recorded, for example, by an analog device, are called *continuous time series*. On the other hand, data observed at certain intervals of time, such as the atmospheric pressure measured hourly, are called *discrete time series* [1].

A time series consisting of a single observation at each time point is called *univariate time series*, whereas, time series obtained by simultaneously recording of two or more phenomena are called *multivariate time series*. A discrete time series $(X(n))$ collected at equally spaced intervals and univariate are of interest in our research. The crude oil prices collected daily, the annual air passengers demand, the monthly sunspot numbers, the weekly fishery stock, are just a few examples of discrete time series.

Relevant historic data are provided by numerous public and private institutions for supporting planning and management tasks. The National Oceanic and Atmospheric Administration of Unite States (NOOA) makes available a list of time series relating to surface pressure, sea surface temperature, and other climate variables in [2]. One objective of NOOA is to assess about current and future states of the climate system that identify potential impacts and inform science, service, and stewardship decisions. Rob Hyndman's Time Series Data Library contains over 800 time series organized by subject as well as pointers to other sources of time series available on the Internet. Finance, transport, sales, hydrology, meteorology, health, and industry are some categories that are freely available in [3]. The Energy Information Administration of U.S. (EIA) makes it available thousands of data sets of quantities, price, consumption, stock, among others about electricity, petroleum, coal, and energy in [4]. Economic indicators as GDP, GNI, exports of goods and services, foreign investment, inflation, consumer prices, and more indicators are available in the The World Bank site in [5].

Most of natural events are impossible to predict deterministically. Those events which evolve in time according with probabilistic laws are *stochastic processes*. An

© Springer International Publishing AG, part of Springer Nature 2018
L. M. Barba Maggi, *Multiscale Forecasting Models*,
https://doi.org/10.1007/978-3-319-94992-5_1

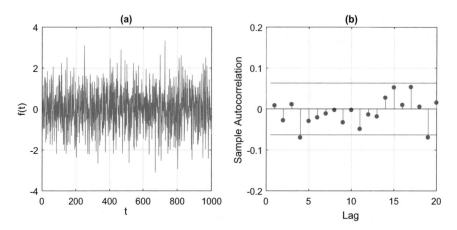

Fig. 1.1 (**a**) Gaussian White Noise (**b**) Autocorrelation Function

stochastic process is non-deterministics and consequently random. The sequence of historical observations is in fact a *sample realization* of the stochastic process that produced it [6].

A completely random process presents zero mean and variance without the presence of trend or seasonal components. This type of process is called i.i.d. noise, in which the observations are simply independent and identically distributed random variables [7]. A noise sequence is a white noise (WN) sequence if the mean is zero, the variance is fixed, and the elements of the sequence are uncorrelated. A Gaussian WN implies a normal distribution, in this case the WN process is an i.i.d. white noise. Figure 1.1 shows a simulation of Gaussian WN signal generated by software and the correspondent plot of the autocorrelation function (ACF).

A deterministic signal is assumed to vary much more slowly than a random process. This type of signal will not depend of random processes. A deterministic signal can be described by sets of analytical expressions which are valid for past, present and future times. This type of signals consist of two or more regions where clearly different phenomena are occurring [8]. Figure 1.2a illustrates a deterministic signal, whereas Fig. 1.2b illustrates the difference of the signal along the first non-singleton dimension.

A time series is stationary when its statistical properties do not vary with the particular value of time. A stationary process remains in *statistical equilibrium* with probabilistic properties that do not change over time, in particular varying about a fixed constant mean and with constant variance [9].

Random processes whose statistical properties do not vary with the particular value of time are much more amenable to analysis than those whose statistical properties do. A nonstationary process at difference of a stationary can be defined as one whose statistical parameters, mean and variance, vary over time. Most *naturally generated* signals are nonstationary, in that the parameters of the system

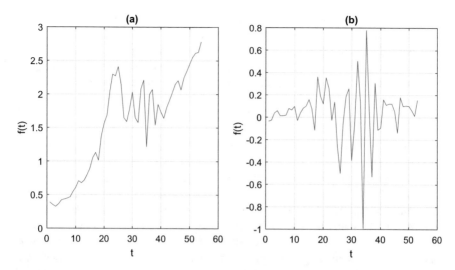

Fig. 1.2 (**a**) A deterministic signal (**b**) The irregular component of the signal extracted via difference

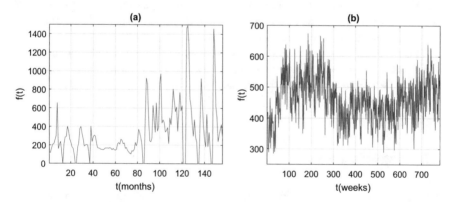

Fig. 1.3 Nonstationary time series (**a**) Shrimp catches in Guayaquil gulf (Tm) (**b**) Number of traffic accidents in Santiago

that generate the signals, and the environments in which the signals propagate, change with time and/or space [10].

Some researchers consider that natural processes are inherently nonstationary, although apparent nonstationarity in a given time series may constitute only a local fluctuation of a process that is in fact stationary on a longer time scale [6] or viceversa. Figure 1.3a, b illustrates time series coming from real world, both present nonstationary features.

The nonstationary processes analysis could be an unsatisfactory task, therefore diverse techniques are implemented to convert them in stationary. The conventional way begins with the identification of components incrusted in a time series $X(n)$.

That process often is critical, in the most complete case, the components of a time series are trend $(T(n))$, cyclical fluctuation $(C(n))$, seasonal component $(S(n))$ and irregular component $(I(n))$ [11]. Decomposition models are typically additive or multiplicative, it is dependent of the amplitude of both the seasonal and irregular variations. An additive decomposition model may be written as follows:

$$X(n) = T(n) + C(n) + S(n) + I(n). \tag{1.1}$$

Trend may be defined as *long-term change in the mean level* [12]. A plot of successive observations of the analyzed process often shows the trend which may be linear, asymptotic or exponential. *Seasonal fluctuations* are exhibited as *high variation* in one part of the observed time period and *low variation* in the remaining time period. For instance rainfall is mainly high in winter but lower in summer, consequently the rainfall series has seasonal variation. On the other hand in commerce field, seasonal variation occurs because of seasonal demand. *Cyclic fluctuations* are related to those variation that occur over a *very long time period*. It is advisable to pay attention in the data collection period, cyclic fluctuations could not be identified when it is lower than their duration. Examples of cyclic fluctuations are El Niño Southern Oscillation, ENSO (which occurs on average every 2–7 years according to [2]), business cycles (which duration depends of economic and technological variables fully explained in [13]), among others.

The *Irregular component* is part of the series that cannot be attributed to any of the other three components. It is caused by short-term unanticipated and non-recurring factors such as strikes, power failures, variations in the weather and so on, and is totally unpredictable [14].

1.2 Analysis Tools

Analysis tools have been implemented to determine stationary or nonstationary characteristics in a time series. The processes can be defined by its mean, variance, and autocorrelation. Once the process is characterized, the model can be selected. The Box-Jenkins methodology suggest to apply differentiation to convert the process in stationary and then to make use of a conventional model.

In certain situations, it may be difficult to ascertain whether or not a given series is nonstationary. This is because there is often no sharp distinction between stationarity and nonstationarity when the nonstationary boundary is nearby. In the Box-Jenkins methodology [9], the Autocorrelation Function (ACF) is often used for identifying, selecting, and assessing conditional mean models (for discrete, univariate time series data).

The ACF is dimensionless, that is, independent of the scale of measurement of the time series. A positive correlation indicates that large current values correspond to large values; whereas a negative correlation at the specified lag indicates that small current values correspond to large values. The absolute value of the correlation is a measure of the strength of the association, values near of ± 1 indicate strong correlation. The confidence limits are provided to display when ACF is significantly different from zero, suggesting that the lags having values outside these limits should be considered to have significant correlation.

The autocorrelation (ρ) between values x_i and x_{i+k}, separated by k intervals of time (lags) is

$$\rho_X(k) = \frac{\gamma_x(k)}{\gamma_x(0)}, \tag{1.2a}$$

$$\gamma_X(k) = \frac{\sum_{i=1}^{N-k}(x_i - \mu_X)(x_{i+k} - \mu_X)}{N-k}, \tag{1.2b}$$

$$\gamma_X(0) = \frac{\sum_{i=1}^{N}(x_i - \mu_X)^2}{N-K}, \tag{1.2c}$$

where μ_X is the average of the N observations. $\gamma_X(k)$ is the covariance of $x(i)$ and $x(i+k)$. $\gamma_x(0)$ is the autocovariance of X at zero lag.

On the other hand, the PACF in general is a conditional correlation which is computed between two variables under the assumption that the values of some other set of variables is known.

Some researchers have used the ACF and PACF information to select the appropriate ARIMA model. Soni et al. shown that the autoregressive and moving average terms of some stationary time series related with aerosols over the Gangetic Himalayan region were determined by examining the patterns of the graphs of ACFs and PACF [15]. Hassan points that for autoregressive processes ARIMA(P,0,0), the ACF declines exponentially and the PACF spikes on the first P lag. By contrast, for moving average processes ARIMA(0,0,Q), the ACF spikes on the first Q lags and the PACF declines exponentially [16].

In general, the plots of ACF and PACF as functions of lag k facilitates the inspection. By example if the shape decays gradually to zero an AR model is considered and the PACF plot is used to identify the order of the autoregressive model. Conversely, for a moving average (MA) process, the shape cuts off after a few lags, but the sample PACF decays gradually. If both the ACF and PACF decay gradually, an ARMA model is considered.

If the ACF plot shows all zero or close to zero, the signal is essentially random. High values at fixed intervals include seasonal autoregressive term. If no decay to zero Series is nonstationary.

Statistical tests are tools to assess about an stationary or nonstationary time series. The test is performed to ascertain if the mean and variance of a random process are constant. The statistical tests allow to accept or reject a hypotheses which formulates that the process has constant mean and variance with a certain level of confidence.

KPSS due to Kwiatkowski et al. is a unit root test which assesses about nonstationary processes [17]. The null hypothesis of trend stationary corresponds to the hypothesis that the variance of the random walk equals zero around a confidence level. In KPSS statistic, the time series $X(n)$ is decomposed into the sum of a deterministic trend $(T(n))$, a random walk $(R(n))$ composed by $R(n-1)$ and i.i.d. processes, and a stationary error $(\varepsilon(n))$ obtained as residuals from a regression of $X(n))$, this is as follows:

$$X(n) = T(n) + R(n) + \varepsilon(n), \tag{1.3}$$

The null hypothesis defines $\sigma_X^2 = 0$ which means that the series is stationary. The size of the test depends on sample size (N) and on the number of covariance lags (k). In the experiment presented in [17] were considered sample sizes from 30 to 500, and covariance lags of $k = 0$, $k = integer[4(N/100)^{1/4}]$, and $k = integer[12(N/100)^{1/4}]$.

The KPSS test has been implemented in some computational packages. The input arguments are the time series $X(n)$, the number of covariance lags k, and the significance level (generally from 0.01% to 0.1%). The output is 0 or 1. The logical value 0 indicates failure to reject the trend-stationary null hypothesis. The logical value 1 indicates rejection of the trend-stationary null, in favor of the unit root alternative.

ADF is the acronym of Augmented Dickey-Fuller test. ADF test for a unit root assesses the null hypothesis of a unit root [18]. Let N observations $x_1, x_2, \ldots, x_N$, be generated by the model,

$$x_i = \phi x_{i-1} + e_i, \qquad i = 1, \ldots, N, \tag{1.4}$$

where ϕ is a real number, $x_0 = 0$, and e_i is a sequence of independent normal random variables with mean zero and variance.

The time series $X(n)$ converges (as $i \to \infty$) to a stationary time series if $|\phi| < 1$. If $|\phi| = 1$, the time series is not stationary and the variance of x is $i\sigma^2$. The time series with $\phi = 1$ is sometimes called a random walk. If $|\phi| > 1$, the time series is not stationary and the variance of the time series grows exponentially as i increases.

Given the observations $x_1, x_2, \ldots, x_N$, the maximum likelihood estimator of ϕ is the least squares estimator,

$$\hat{\theta} = \left(\sum_{i=1}^{N} x_{i-1}^2 \right)^{-1} \sum_{i=1}^{N} x_i x_{i-1}. \tag{1.5}$$

The ADF test has been implemented in some computational packages. The input arguments are the time series $X(n)$, the covariance lags, and the significance level (generally from 0.01% to 0.1%). The output are 0 or 1, the logical value 0 which indicates failure to reject the unit-root null, or the logical value 1 which indicates rejection of the unit-root null in favor of the trend-stationary alternative.

1.3 Linear Autoregressive Models

Linear time series modelling are commonly associated with a family of linear stochastic models which are referred as ARIMA (Autoregressive Integrated Moving Average) models. ARIMA is also known as Box-Jenkins methods due to the work of George P. Box and Gwilym M. Jenkins published in 1970 [19]. ARIMA is in fact a culmination of the research of many prominent statisticians, starting with the pioneering work of Yule in 1927, who employed an Autoregressive (AR) model of order 2 to model yearly sunspot numbers [6].

Time series analysis by means of linear models are conditioned to deal with stationary data or at least weakly stationary. Stationary models assume that the process remains in *statistical equilibrium* with probabilistic properties that do not change over time, in particular varying about a fixed constant mean level and with constant variance [9], which means that the data are no trend nor seasonality and homoscedastic [20].

Most *naturally generated* signals are nonstationary, in that the parameters of the system that generate the signals, and the environments in which the signals propagate, change with time and/or space [10]. Some researchers consider that natural processes are inherently nonstationary, although apparent nonstationarity in a given time series may constitute only a local fluctuation of a process that is in fact stationary on a longer time scale [6] or viceversa.

There are some methods to convert a nonstationary time series in stationary. ARIMA implements D processes of differentiation, most time series are stationarized taking the first difference ($D = 1$). An ARMA(P, Q) model is the generalization of an ARIMA(P, D, Q) model, where P is the order of the autoregressive part (number of autoregressive terms) and Q is the order of the moving average part (number of lagged forecast errors).

The simplest analysis is performed through an autoregressive model of order P and with i.i.d. innovations ε (having zero mean and at least finite second-order moments) for the data-generating process [21],

$$\widehat{X}(n + 1) = \sum_{i=1}^{P} \alpha_i Z_i + \varepsilon(n + 1), \tag{1.6}$$

where P is the model order, α_i is ith coefficient and Z_i is the ith regressor vector.

An ARMA(P, Q) model is defined with

$$\widehat{X}(n + 1) = \sum_{i=1}^{P} \alpha_i Z_i + \sum_{i=1}^{Q} \beta_i \varepsilon_i + \varepsilon(n + 1), \tag{1.7}$$

where $\widehat{X}(n + 1)$ is the future value of the observed time series, α_i is the i-th coefficient of the AR term Z_i. β_i denotes the ith coefficient of the MA term ε_i

and $\varepsilon(n + 1)$ is a source of randomness which is called white noise. The AR terms Z_i are the columns of the regressor matrix $Z = (Z_1, \ldots, Z_P)$.

Consequently an AR(P) model is defined with

$$\widehat{X}(n + 1) = \sum_{i=1}^{P} \alpha_i Z_i. \tag{1.8}$$

The coefficients of AR and ARMA models are commonly estimated by Least Squares (LS) method and by Maximum Likelihood Estimation (MLE). LS computes the parameters that provide the most accurate description of the data, the sum of square errors computed between observed values and estimated values must be minimized. Whereas MLE is a method to seek the probability distribution that makes the observed data most likely.

Least Squares method is one of the oldest technique of modern statistics which origin is related with the work of Legendre and Gauss [22]. In a standard formulation, a set of pairs of observations x, y, is used to find a function that relates the observations. A linear function is defined to estimate the values of a set of dependent variables Y from the values of independent variables X, as follows:

$$\widehat{Y} = a + bX, \tag{1.9}$$

where a is the intercept and b is the slope of the regression line. If the intercept is zero, the equation is reduced to

$$\widehat{Y} = bX. \tag{1.10}$$

The LS method estimates the b parameters according to the rule that those values must minimize the sum of residual squares computed between the observed values and the predicted values,

$$\varepsilon = \sum_{i=1}^{N}(y_i - \widehat{y}_i)^2, \tag{1.11a}$$

$$= \sum_{i=1}^{N}(y_i - bx_i)^2, \tag{1.11b}$$

where ε is the value to be minimized. Taking the derivative of ε with respect to b, and setting them to zero gives the following equation:

$$\frac{\partial \varepsilon}{\partial b} = \frac{\partial \left(\sum_i (y_i^2 - 2bx_i y_i + b^2 x_i^2) \right)}{\partial b}, \tag{1.12a}$$

$$b = \frac{\sum_i x_i y_i}{\sum_i x_i^2}. \tag{1.12b}$$

Extending the solution to a higher degree polynomial is straightforward. In matrix form, linear models are given by the formula

$$\widehat{X} = \alpha Z, \tag{1.13}$$

where X is a $N \times 1$ vector of results (for univariate time series), α is a $P \times 1$ vector of coefficients (with $P < N$), and Z is the regressor matrix of $N \times P$ dimension, which is named design matrix for the model.

The Moore-Penrose pseudoinverse matrix [23] $X^\dagger$ is computed to obtain the coefficients matrix α, it is

$$\alpha = Z^\dagger X. \tag{1.14}$$

In general, LS estimates tend to differ from MLE estimates, especially for data that are not normally distributed such as proportion correct and response time.

Maximum Likelihood Estimation (MLE) was first introduced by Fisher in 1922, although he first presented the numerical procedure in 1912 [24]. MLE consists of choosing from among the possible values for the parameter, the value which maximizes the probability of obtaining the sample which was obtained [25].

In the practice, from original observations (y_i for $i = 1, \ldots, N$) through an ARIMA(P, D, Q) model is generated a new differentiated series $X(n) = x_1, \ldots, x_N$. The parameters set is defined with $\theta = (\alpha_1, \ldots, \alpha_P, \beta_1, \ldots, \beta_Q, \sigma^2)$. The joint probability density function (PDF) is denoted with

$$f(x_N, x_{N-1}, \ldots, x_1; \theta). \tag{1.15}$$

The *likelihood function* is this joint PDF treated as a function of the parameters θ given the data $X(n)$:

$$L(\theta|X) = f(x_N, x_{N-1}, \ldots, x_1; \theta). \tag{1.16}$$

The maximum likelihood estimator is

$$\widehat{\theta} = \arg\ \max\ L(\theta|X(n)), \qquad \theta \in \Theta, \tag{1.17}$$

where Θ is the parameters space. The term arg max refers to the arguments at which the function output is as large as possible.

For simplifying calculations, it is customary to work with the natural logarithm of L, which is commonly referred to as the *log-likelihood function*. Because the logarithm is a monotonically increasing function, the logarithm of a function achieves its maximum value at the same points as the function itself, and hence the log-likelihood can be used in place of the likelihood in maximum likelihood estimation and related techniques. Finding the maximum of a function often involves taking the derivative of a function and solving for the parameter being maximized, and this is often easier when the function being maximized is a log-likelihood rather than the original likelihood function.

MLE is preferred to LS when the probability density function is known (generally normal) or easy to obtain through computable form. There is a situation in which LS and MLE intersect. This is when observations are independent of one another and are normally distributed with a constant variance. In this case, maximization of the log-likelihood is equivalent to minimization of SSE, and therefore, the same parameter values are obtained under either LS or MLE [26].

ARIMA model have shown several applications with optimum results especially for short term time series forecasting [27]. By instance Ramos et al. performed forecasting of shoes sales for 5 retail series by means of an ARMA model, P and Q orders were chosen from 0 to 5, the lowest average Mean Absolute Percentage Error (MAPE) of 22.3% was obtained for shoes series whereas the highest average MAPE was 185.2% for sandals, both for 12-months ahead forecasting [28].

A time series with seasonal variation about aerosol optical depth was forecasted via multiplicative ARIMA in [29]. Seasonal differencing was performed to make it stationary. ARIMA $(1,0,0) \times (2,1,1)_{12}$ shown the least error with an R^2 of 0.68 for one-step ahead forecasting.

Forecast of wind speed was made in [30]. An ARMA model was implemented once the time series was transformed to stationary. It was observed the deterioration of the accuracy of the forecast as the forecasting period increases.

1.4 Artificial Neural Networks

An Artificial Neural Network (ANN) is a nonlinear technique used conventionally in time series prediction. The first ANN was proposed by McCulloch and Pitts in 1943, it was designed to use connections with a fixed set of weights. In the early 1960s, some researchers among them Rosenblatt, Block, Minsky and Paper developed a learning algorithm for a device known as Perceptron which convergence was warranted if the connections weights could be adjusted. In 1969 Minsky and Papert, determined that the Perceptron was not able to generalize the learning with nonlinear functions. The limitation of the Perceptron, paralyzed for a long time the researching in the field of ANNs.

Between 1980 and 1986, the PDD group consisting of Rumelhart and McCelland, publish the book Parallel Distributed Processing: Explorations in the Microstructure of Cognition [31]. Backpropagation (BP) algorithm was shown in the publication as an algorithm for multilayer ANNs known as Multilayer Perceptron (MLP).

ANN is in general an information processing system that has certain performance characteristics in common with biological neural networks [32]. As Fausett an ANN is based on some assumptions:

- Information processing occurs at many simple elements called neurons.
- Signals are passed between neurons over connection links.
- Each connection link has associated weight, which, in a typical neural net, multiplies the signal transmitted.
- Each neuron applies an activation function (which usually is nonlinear) to its net input (sum of weighted input signals) to determine its output signal.

The performance of an ANN depends of different elements, such as, structure, activation function and learning algorithm. Classification, prediction, and images recognition, are common tasks of an ANN. The power of an ANN approach lies not necessarily in the elegance of the particular solution, but rather in the generality of the network to find its own solution to particular problems, given only examples of the desired behavior [33].

The common structure of an ANN has three layers, input, hidden, and output. The presence of a hidden layer together with a nonlinear activation function, gives it the ability to solve many more problems than can be solved by a net with only input and output layers [32].

An autoregressive ANN of three layers uses the lagged terms z_i at the input layer, they are weighted with respect to the hidden layer, and at the output of the hidden layer is applied the activation function. At the output layer is obtained the estimated value $\widehat{X}(n + 1)$,

$$\widehat{X}(n + 1) = \sum_{j=1}^{Q} b_j Y_{Hj}, \tag{1.18a}$$

$$Y_{Hj} = f\left(\sum_{i=1}^{P} w_{ji} z_i\right), \tag{1.18b}$$

where $Z_1, \ldots, Z_P$ are the input neurons, and $w_{11}, \ldots, w_{P1}, \ldots, w_{PQ}$ are the nonlinear weights of the connections between the inputs and the hidden neurons. Whereas $b_1, \ldots, b_Q$ are the weights of the connections between the hidden neurons and the output (under the assumption that there is an unique output).

The common activation function is sigmoid, which is expressed with

$$f(x) = \frac{1}{1 + e^{-x}} \tag{1.19}$$

The ANN for an univariate process is denoted with $ANN(P, Q, 1)$, with P inputs, Q hidden nodes, and 1 output. The parameters w_{ij} and b_j are updated during the training via learning algorithm. At each iteration the weights are adapted to reach the minimization of the errors.

The learning algorithms are gradient-based and gradient-free (or derivative-free). Algorithms of first order are those where the first derivative of the objective function is computed to adapt the ANN weights, whereas algorithms of second order are those that compute the second derivative.

Backpropagation is a gradient-based also known as steepest-descent algorithm which repeatedly adjusts the weights of the connections in the network so as to minimize a measure of the difference between the actual output vector [34]. Standard steepest descent is denoted as follows,

$$\Delta \omega^t = l_r \frac{\partial E^t}{\partial \omega^t}, \tag{1.20a}$$

$$\omega^{t+1} = \omega^t - \Delta \omega^t, \tag{1.20b}$$

where $\Delta \omega_t$ is called delta rule, and determines the amount of weight update based on the gradient direction along with a step size. ω is the matrix of weights and bias, t is the number of epoch (repetition), l_r is the learning rate (in standard steepest descent BP, l_r is constant). E is the *Performance Function* which is arbitrary (E generally has a quadratic form, by example Mean Square Error MSE). The ANN is trained until the stopping criteria is reached; stop criteria might be the number of epochs, the maximum iteration time, the minimum level of error performance, or the performance gradient falls below the minimum gradient predefined.

Backpropagation with adaptive learning and momentum coefficient is an improved version of standard BP. The momentum coefficient and the learning rate are adjusted at each iteration to reduce the training time [35],

$$\Delta \omega^t = m_c \partial \omega^{t-1} + l_r m_c \frac{\partial E^t}{\partial \omega^t}, \tag{1.21}$$

where $\partial \omega^{t-1}$ is the previous change to the weights (or bias), and m_c is the momentum coefficient. The training stops when any of the criteria that where defined for standard BP is achieved.

The main disadvantage of conventional gradient-descent method is the slow convergence. Whereas for the improved BP version, the momentum parameter is equally a problem dependent as the learning rate, and that no general improvement can be accomplished [36].

However, since the algorithm employs the steepest descent technique to adjust the network weights, it suffers from a slow convergence rate and often produces suboptimal solutions, which are the main drawbacks of BP [37]. Steepest descent method presents accelerated convergence through adaptive learning rate and momentum factor [38, 39].

Resilient Backpropagation (RPROP) was proposed by Riedmiller and Braun in 1993 [36]. RPROP outperforms the BP limitations through a direct adaptation of the weight step based on the local gradient information [40]. The update rule depends only of the sign of the partial derivative of the arbitrary error regards to each weight of the ANN. The individual step size Δ^t is updated for each weight as follows:

$$\Delta^t = \begin{cases} \eta^+ * \Delta^{t-1} & \text{if } \frac{\partial E^{t-1}}{\partial \omega} * \frac{\partial E^{t-1}}{\partial \omega} > 0 \\ \eta^- * \Delta^{t-1} & \text{if } \frac{\partial E^{t-1}}{\partial \omega} * \frac{\partial E^{t-1}}{\partial \omega} < 0 \\ \Delta^{t-1} & \text{else} \end{cases} \qquad (1.22)$$

where $0 < \eta^- < 1 < \eta^+$.

Every time the partial derivative of the weights ω changes its sign, which indicates that the last update was too big and the algorithm has jumped over a local minimum, the updated value Δ is decreased by the factor η^-. If the derivative retains its sign, the updated value is slightly increased in order to accelerate convergence in shallow regions.

Once the update value for each weight is adapted, the weight update itself follows a very simple rule: if the derivative is positive (increasing error), the weight is decreased by its update value, if the derivative is negative, the update value is added:

$$\Delta\omega^t = \begin{cases} -\Delta^{t-1} & \text{if } \frac{\partial E^{t-1}}{\partial \omega} > 0 \\ +\Delta^{t-1} & \text{if } \frac{\partial E^{t-1}}{\partial \omega} < 0 \\ 0 & \text{else} \end{cases} \qquad (1.23)$$

Marquardt in 1963 published a method called *maximum neighborhood* to perform an optimum interpolation between the Taylor series method and the gradient method to represent a nonlinear model [41]. The Marquardt studies were based on the previous work of Levenberg in 1944 [42].

Levenberg-Marquardt (LM) is a second order algorithm that outperforms the accuracy of the gradient-based methods for a widely variety of problems [43–45].

The scalar u controls the LM behavior. If u increases the value, the algorithm works as the steepest descent algorithm with low learning rate; whereas if u decreases the value until zero, the algorithm works as the Gauss-Newton method [46]. The weights of the ANN connections are updated with:

$$\omega^{t+1} = \omega^t - \Delta\omega^t \qquad (1.24a)$$

$$\Delta\omega^t = \left[J^T(\omega^t) J(\omega^t) + u^t I \right]^{-1} J^T(\omega^t) \varepsilon(\omega^t), \qquad (1.24b)$$

where $\Delta\omega^t$ is the weight increment, J is the Jacobian matrix, T is used to transposed matrix, I is the identity matrix, and ε is the error vector.

The Jacobian matrix is created with the computation of the derivatives of the errors, instead of the derivatives of the squared errors,

$$
J = \begin{bmatrix}
\dfrac{\partial e_{1,1}}{\partial w_{1,1}} & \dfrac{\partial e_{1,1}}{\partial w_{1,2}} & \cdots & \dfrac{\partial e_{1,1}}{\partial w_{P,H}} & \dfrac{\partial e_{1,1}}{\partial b_1} & \cdots \\[2mm]
\dfrac{\partial e_{2,1}}{\partial w_{1,1}} & \dfrac{\partial e_{2,1}}{\partial w_{1,2}} & \cdots & \dfrac{\partial e_{2,1}}{\partial w_{P,H}} & \dfrac{\partial e_{2,1}}{\partial b_1} & \cdots \\[2mm]
\vdots & \vdots & & \vdots & \vdots & \\[2mm]
\dfrac{\partial e_{P,1}}{\partial w_{1,1}} & \dfrac{\partial e_{P,1}}{\partial w_{1,2}} & \cdots & \dfrac{\partial e_{P,1}}{\partial w_{P,H}} & \dfrac{\partial e_{P,1}}{\partial b_1} & \cdots \\[2mm]
\dfrac{\partial e_{1,2}}{\partial w_{1,1}} & \dfrac{\partial e_{1,2}}{\partial w_{1,2}} & \cdots & \dfrac{\partial e_{1,2}}{\partial w_{P,H}} & \dfrac{\partial e_{1,2}}{\partial b_1} & \cdots \\[2mm]
\vdots & \vdots & & \vdots & \vdots &
\end{bmatrix}
\tag{1.25}
$$

The activation function is an important element in an ANN used to reach excitation at hidden layer or at output layer. The choice of an activation function depends on how it is required to represent the data at the output. A sigmoid activation function increases monotonically on real numbers and have finite limits in the whole interval, a requirement to have a positive derivative at every real point. Sigmoid activation functions and Levenberg-Marquardt are commonly observed in forecasting models based on ANNs in diverse areas, for instance in water transportation [47], customers demand [48], stock price [49], among others.

Comparisons among linear forecasting models and nonlinear forecasting models, specifically with Artificial Neural Networks, have shown that ANNs are more accurate [50, 51]. However some authors have argued that evidence of superior forecasting performance of nonlinear models with respect to linear forecasting models is patchy, besides one factor that has probably retarded the widespread reporting of nonlinear forecasts is that up to that time it was not possible to obtain closed-form analytical expressions for multi-step-ahead forecasts [52].

1.5   Forecasting Accuracy Measures

Accuracy measures have been used to evaluate the performance of forecasting methods. Forecasting by means of univariate time series is only dependent of the present and past values of the series being estimated. A common procedure is to split the time series into two portions, the first is called training sample and the second portion is called testing sample. The training sample is used to fit the model and the testing sample is used to check the estimation accuracy (*goodness of fit*).

The evolution of measures of forecast accuracy and evaluation can be seen through the measures used to evaluate methods in the major comparative studies that have been undertaken. Makridakis et al. in 1982 [53] used common measures included MAPE (Mean Absolute Percentage Error) and MSE (Mean Square Error). However, the MSE is not appropriate for comparison between series as it is scale dependent; whereas MAPE also has problems when the series has values close to (or equal to) zero [52].

In general the use of absolute errors and square errors avoid the errors canceling of opposite sign. This is the case of Mean Absolute Percentage Error ($MAPE$),

Root Mean Square Error ($RMSE$), normalized Root Mean Square Error ($nRMSE$), General Cross Validation (GCV) and Coefficient of Determination (R^2).

$$MAPE = \frac{1}{N} \sum_{i=1}^{N} \left| \frac{x_i - \widehat{x_i}}{x_i} \right|. \tag{1.26}$$

$$RMSE = \sqrt{\frac{1}{N} \sum_{i=1}^{N} (x_i - \widehat{x_i})^2}. \tag{1.27}$$

$$nRMSE = \frac{RMSE}{\bar{x}}. \tag{1.28}$$

$$GCV = \frac{RMSE}{(1 - P/N)^2}. \tag{1.29}$$

$$R^2 = 1 - \frac{\sigma^2(X(n) - \widehat{X}(n))}{\sigma^2(X(n))}. \tag{1.30}$$

Further, some efficiency criteria make emphasis on sensitivity to significant overfitting or underfitting when the observed values are very large or very small, respectively. Krause et al. [54], shown the evaluation of nine different efficiency measures for the evaluation of model performance in the hydrologic field. The study shown that modified forms of Nash-Sutcliffe Efficiency ($mNSE$) and Index of Agreement (mIA) increase the sensitivity of efficiency measures to low flow conditions.

The modified Nash-Suctlife Efficiency ($mNSE$)is based on the Sum of Absolute Error (SAE) and the Sum of Absolute Deviation (SAD); the formulas are shown below,

$$SAE = \sum_{i=1}^{N} |x_i - \widehat{x_i}|, \tag{1.31a}$$

$$SAD = \sum_{i=1}^{N} |x_i - \mu_X|, \tag{1.31b}$$

$$mNSE = 1 - \frac{SAE}{SAD}. \tag{1.32}$$

The modified Index of Agreement (mIA) [55], as $mNSE$, overcomes the possible oversensitivity to extreme values induced by metrics based on square computation, and increase the possible sensitivity for lower values.

$$mIA = 1 - \frac{SAE}{2SAD}, \quad if \quad SAE \leq 2SAD, \tag{1.33a}$$

$$mIA = \frac{2SAD}{SAE} - 1, \quad if \quad SAE > 2SAD, \tag{1.33b}$$

Measure mIA is monotonically and functionally related, but the use of $2SAD$ balances the number of deviations evaluated within the numerator and within the denominator of the factional part.

An additional measure commonly applied in forecasting is the Relative Error (RE). Scale independence, outlier independence, sensitivity, and stability are some properties of RE [56].

$$RE = \frac{X(n) - \widehat{X}(n)}{X(n)}. \tag{1.34}$$

1.6   Empirical Application: Traffic Accidents Forecasting

The traffic accidents occurrence is a matter of impact in the society, therefore a problematic of priority public attention. The Chilean National Traffic Safety Commission (CONASET), periodically reports a high rate of sinister on roads, in Valparaíso from year 2003 to 2012 were registered 28,595 injured people. The accuracy in the projections enables the intervention of government agencies in terms of prevention. In this empirical application is used a time series of people with injuries in traffic accidents, the data were collected in Valparaíso.

The observed time series is preprocessed via smoothed values given by 3-points Moving Average. Whereas one-step ahead forecasting is implemented through 3 commonly accepted models, ARIMA, (MA-ARIMA), an ANN supported by RPROP algorithm (MA-ANN-RPROP), and an ANN supported by Levenberg-Marquardt (MA-ANN-LM). The forecasting strategy is illustrated in Fig. 1.4, while the raw time series is presented in Fig. 1.5.

1.6.1   Smoothing Data via Moving Average

Moving Average is a conventional smoothing strategy used in linear filtering to identify or extract the trend from a time series. MA is a mean of a constant number of observations that can be used to describe a series that does not exhibit a trend [57]. When 3-point MA is applied over a time series of length n, the $n - 2$ elements of the smoothed series are computed with

Fig. 1.4 Conventional
forecasting based on Moving
Average, AR, and ANNs

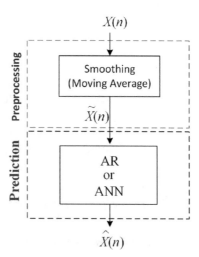

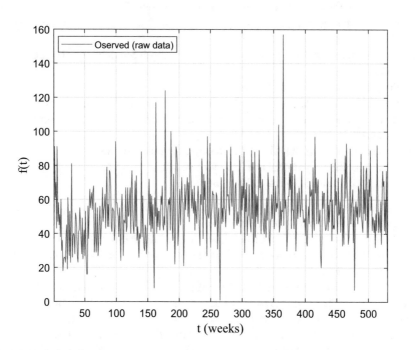

Fig. 1.5 Raw data of injured in traffic accidents, Valparaíso

$$\widetilde{x}_k = \frac{1}{3} \sum_{i=k-1}^{k+1} x_i \tag{1.35}$$

where $\widetilde{x}_k$ is the k-th smoothed element, for $k = 2, \ldots, N - 1$, x_i is each observed element of original time series, terms $\widetilde{x}_1$ and $\widetilde{x}_N$ has the same value of x_1 and x_N respectively.

The data was split in two groups, training sample and testing sample, 453 record are used for training and 78 records for testing.

1.6.2 One-Step Ahead Forecasting Based on AR Model

The solution is developed with the following sequential steps:

1. Normalize the time series values (number of people injured in traffic accidents), in this case is computed the ratio between each observed value and the maximum value.
2. Apply a moving average filter (MA-3), to smooth the data and the influence of outliers.
3. Split the data en training and testing.
4. Definition of the AR model of P variable order (to identify the effective P value, will be used the GCV metric), at time create a matrix of lagged (time-shifted) series, using P lags.
5. Implementation of the AR model through the training sample.
6. Obtaining the polynomial coefficients through LS method.
7. Model validating with the testing sample.

The smoothed time series is presented in Fig. 1.6, the plot shows that the observed time series is fit with a smoothing spline.

From Fig. 1.7, the GCV metric show that at Lag 26, the lowest GCV value is reached; therefore the effective order is set to $P = 26$. The AR model is denoted with MA-AR(26).

The results of the forecasting via testing sample is presented in Fig. 1.8 and Table 1.1. The observed values versus the estimated values are illustrated in Fig. 1.8, reaching a good accuracy, while the relative error is presented in Fig. 1.9.

The results presented in Table 1.1 show that the model MA-AR(26) reaches a *RMSE* of 0.0206, a *MAPE* of 5.8%, a *GCV* of 0.03, and 90.1% of predicted points present a relative error lower than the $\pm10\%$.

The evaluation of serial correlation of models errors is computed with the Autocorrelation Function. The ACF of the errors obtained via MA-AR were shown in Fig. 1.9. The plot presents all coefficients inside of the confidence limit at 95%.

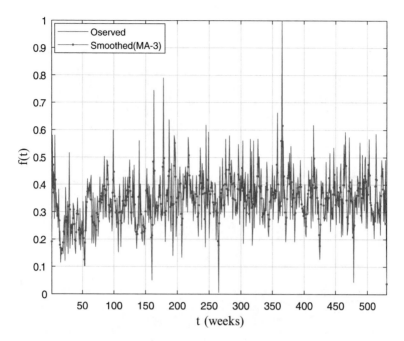

Fig. 1.6 Observed and smoothed time series

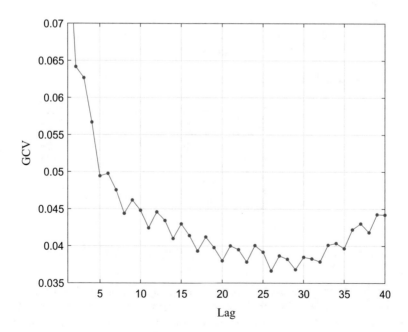

Fig. 1.7 Lags vs GCV to identify the order P for AR model

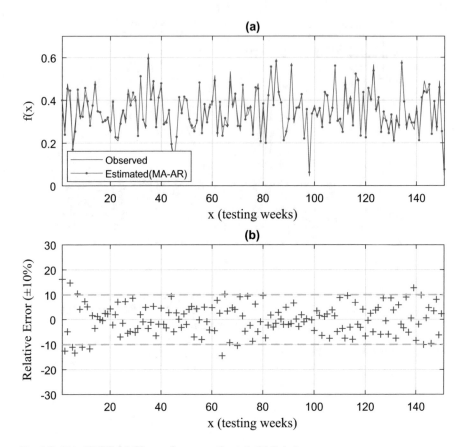

Fig. 1.8 MA-AR(26) (**a**) Observed versus estimated, (**b**) Relative error

Table 1.1 Forecasting results

	MA-AR	MA-ANN-RPROP	MA-ANN-LM
MAE	0.016	0.031	0.017
$RMSE$	0.021	0.039	0.022
$MAPE$	5.79%	11.84%	6.76%
GCV	0.03	0.057	0.032
$RE\pm 10\%$	90.1%	–	90.7%
$RE\pm 15\%$	–	87.0%	

1.6.3   One-Step Ahead Forecasting Based on ANN-RPROP

The raw time series smoothed by moving average filter of order 3, are the inputs of the forecasting model (illustrated in Fig. 1.4). The number of inputs for the ANN-RPROP is set using the previous configuration of AR $P = 26$. The number of hidden neurons is set through the evaluation of an increased number from 0 to 20.

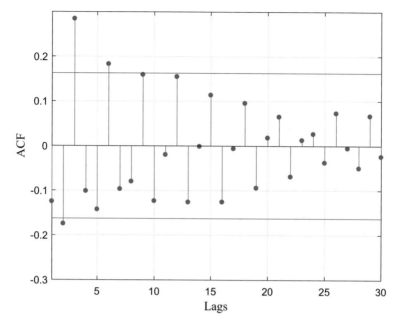

Fig. 1.9 Residuals ACF for MA-AR(26)

Figure 1.10 shows that the best GCV accuracy is reached with $Nh = 7$. The nonlinear model is then denoted with $ANN(26, 7, 1)$.

Once the optimum number of hidden numbers is found, the ANN-RPROP is implemented with the training sample. The process was run 30 times, and the best result was obtained in the run 26 as Fig. 1.11. The *GCV* metric is used to identify the best run.

The process was run 30 times using the testing sample, and the best result was reached in the run 26 as Fig. 1.11. The *RMSE* metric is used to identify the best run. One-step forecasting results are show in Fig. 1.12a, b, and Table 1.1 good accuracy is observed.

The evaluation of serial correlation of ANN-RPROP model errors is computed with the Autocorrelation Function. The ACF of the errors obtained via MA-ANN-RPROP is shown in Fig. 1.13. The plot presents that all ACF values are lower than 95% of confidence limit, consequently, it is concluded that there is no serial correlation in the errors for MA-ANN-RPROP(26,7,1) model; consequently, the proposed model explains efficiently the variability of the process.

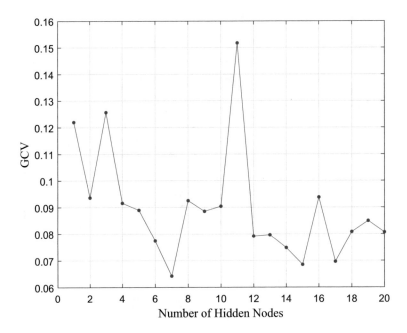

Fig. 1.10 Hidden nodes vs. GCV metric to identify the best accuracy

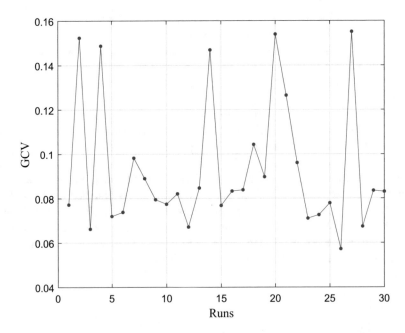

Fig. 1.11 MA-ANN-RPROP(26,7,1): Run vs. GCV for 30 iterations

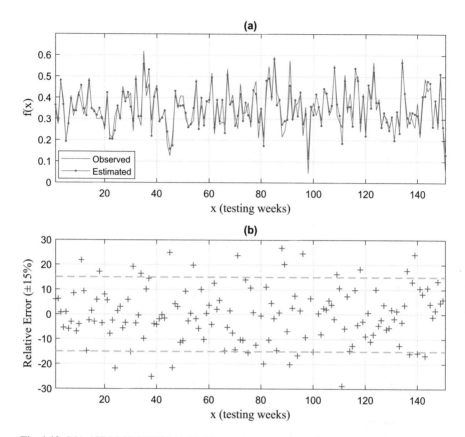

Fig. 1.12 MA-ANN-RPROP(26,7,1) (**a**) Observed versus estimated vales, (**b**) Relative error

1.6.4   One-Step Ahead Forecasting Based on ANN-LM

The number of inputs of the ANN-LM was set using the previous configuration of ANN-RPROP. The number of hidden neurons was found through a similar process as previous model, as shows Fig. 1.14, the best accuracy was reached with 2 hidden nodes, then, this nonlinear model is denoted with $ANN(26, 2, 1)$. The process was run 30 times, as shown in Fig. 1.15. The best results are reported, the evaluation executed with the testing sample is presented in Fig. 1.16a, b, and Table 1.1.

The evaluation of serial correlation of models errors is computed with the Autocorrelation Function. The ACF of the errors obtained via MA-ANN-LM were shown in Fig. 1.17. The plot presents all coefficients inside of the confidence limit at 95%.

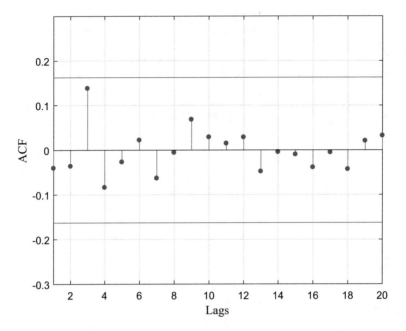

Fig. 1.13   Residual ACF for MA-ANN-RPROP(26,7,1)

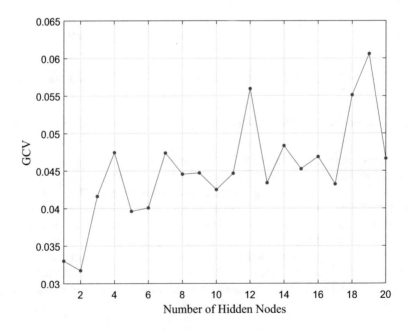

Fig. 1.14   Hidden nodes vs. GCV metric to identify the best accuracy

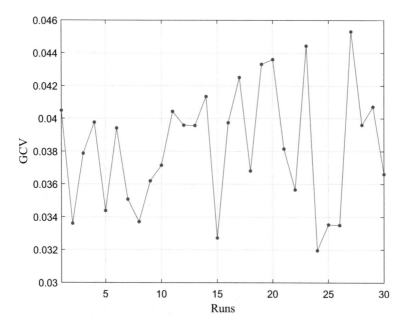

Fig. 1.15 MA-ANN-LM(26,2,1): Run vs GCV for 30 iterations

1.7 Chapter Conclusions

Time series analysis and forecasting are two big research tasks that attract the attention of several sectors.

ARIMA implements D processes of differentiation for transforming a non-stationary time series in stationary, most time series are stationarized taking the first difference ($D = 1$). An ARMA(P, Q) model is the generalization of an ARIMA(P, D, Q) model, where P is the order of the autoregressive part (number of autoregressive terms) and Q is the order of the moving average part (number of lagged forecast errors). The AR(P) model only uses lagged (time-shifted) points.

One popular method to preprocess noise data, is the moving average filter, its application obtains smoothed data and drives the influence of outliers.

On the other hand nonlinear relationships are often modelled by nonlinear models. An ANN is a popular method of artificial intelligence commonly implemented to solve regression problems. An ANN is free statistical parametric assumptions, which means that it deals with nonstationary and/or nonlinear time series due to their nonlinear nature.

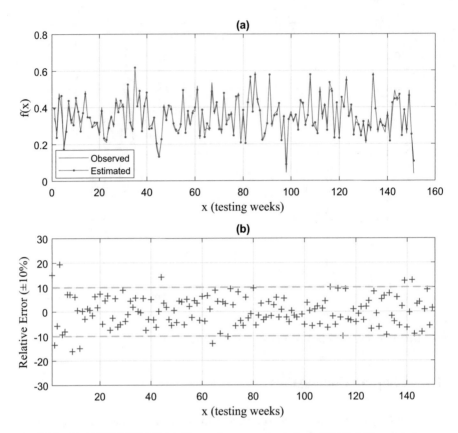

Fig. 1.16 MA-ANN-LM(26,2,1) (**a**) Observed versus estimated vales, (**b**) Relative error

In the empirical application section it was presented one-step ahead forecasting in traffic accidents domain, AR, an ANN based on RPROP, and an ANN based on Levenberg-Marquardt were presented. The results shown that MA-AR, and MA-ANN-LM reach the best accuracy, however MA-ANN-LM and MA-ANN-RPROP present all residual autocorrelations near to zero, inside of confidence bounds, at difference of MA-AR which presents 3 points out of the confidence bounds.

In the next chapter will be presented the HSVD strategy to improve the accuracy of linear and nonlinear autoregressive models.

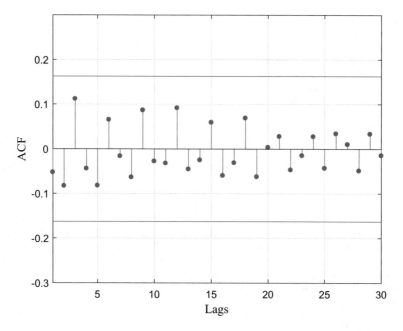

Fig. 1.17 Residual ACF for MA-ANN-LM(26,2,1)

References

1. Kitagawa G (2010) Introduction to time series modeling. Chapman & Hall/CRC Press, Boca Raton; Taylor & Francis Group, Abingdon, pp 6–8
2. NOOA (2016) National Oceanic and Atmospheric Administration. http://www.esrl.noaa.gov/. Visited 01 Aug 2016
3. TSDL (2016) Time Series Data Library. https://datamarket.com/data/list/. Visited 01 Aug 2016
4. EIA (2016) Energy Information Administration. https://www.eia.gov/tools/models/timeseries. cfm/. Visited 01 Aug 2016
5. TWB (2016) The World Bank Open Data. http://databank.worldbank.org/data/home.aspx/. Visited 01 Aug 2016
6. Hipel K, McLeod A (1994) Time series modelling of water resources and environmental systems. Elsevier, New York, p 65
7. Brockwell PJ, Davis RA (2002) Introduction to time series and forecasting, 2nd edn. Springer, Berlin, pp 179–219
8. Hardin JC (1986) Introduction to time series analysis. Langley Research Center, Hampton
9. Box G, Jenkins GM, Reinsel GC, Ljung GM (2009) Time series analysis: forecasting and control, 5th edn. Wiley, Hoboken
10. Vaseghi SV (2006) Advanced digital signal processing and noise reduction, 3rd edn. Wiley, Hoboken, p 138
11. Davey AM, Flores BE (1993) Identification of seasonality in time series: a note. Math Comput Model 18(6):73–81
12. Chatfield C (2003) The analysis of time series: an introduction, 6th edn. Taylor & Francis Group LLC, Abingdon
13. Schumpeter JA (1934) Business cycles: a theoretical, historical, and statistical analysis of the capitalist process, vol 1. McGraw-Hill Book Company, New York

14. Honert RV (1999) Intermediate statistical methods for business and economics, 2nd edn. UCT Press, Cape Town
15. Soni K, Kapoor S, Singh-Parmar K, Kaskaoutis DG (2014) Statistical analysis of aerosols over the Gangetic Himalayan region using ARIMA model based on long-term MODIS observations. Atmos Res 149:174–192
16. Hassan J (2014) ARIMA and regression models for prediction of daily and monthly clearness index. Renew Energy 68:421–427
17. Kwiatkowski D, Phillips P, Schmidt P, Shin Y (1992) Testing the null hypothesis of stationarity against the alternative of a unit root. J Econ 54(1):159–178
18. Dickey DA, Fuller WA (1979) Distribution of the estimators for autoregressive time series with a unit root. J Am Stat Assoc 74(366):427–431
19. Box G, Jenkins GM, Reinsel GC (2008) Time series analysis: forecasting and control, 4th edn. Wiley, Hoboken
20. Diagne M, David M, Lauret P, Boland J, Schmutz N (2013) Review of solar irradiance forecasting methods and a proposition for small-scale insular grids. Renew Sust Energ Rev 27:65–76
21. Kreiss JP, Lahiri SN (2012) Bootstrap methods for time series. In: Time series analysis: methods and applications. Handbook of statistics, vol 30. Elsevier, New York, pp 3–26
22. Plackett RL (1972) Studies in the history of probability and statistics. XXIX: the discovery of the method of least squares. Biometrika 59(2):239–251
23. Golub GH, Van Loan CF (1996) Matrix computations. The Johns Hopkins University Press, Baltimore, pp 48–75
24. Aldrich J (1997) R. A. Fisher and the making of maximum likelihood 1912–1922. Stat Sci 12(3):162–176
25. Beck JV, Arnold KJ (1934) Parameter estimation in engineering and science. Wiley, Hoboken
26. Myung IJ (2003) Tutorial on maximum likelihood estimation. J Math Psychol 47(1):90–100
27. Price BA (1992) Business forecasting methods. Int J Forecast 7(4):535–536
28. Ramos P, Santos N, Rebelo R (2015) Performance of state space and ARIMA models for consumer retail sales forecasting. Robot Comput Integr Manuf 34:151–163
29. Taneja K, Ahmad S, Ahmad K, Attri SD (2016) Time series analysis of aerosol optical depth over New Delhi using Box - Jenkins ARIMA modeling approach. Atmos Pollut Res 7:585–596
30. Torres JL, García A, De Blas M, De Francisco A (2005) Forecast of hourly average wind speed with ARIMA models in Navarre (Spain). Sol Energy 79(1):65–77
31. McClelland JL, Rumelhart DE (1986) A handbook of models, programs, and exercises. MIT Press, Cambridge
32. Fausett L (1993) Fundamentals of neural networks, architectures, algorithms, and applications. Prentice Hall, Upper Saddle River, pp 3–20
33. Freeman JA, Skapura DM (1991) Neural networks, algorithms, applications, and programming techniques. Addison-Wesley, Boston
34. Rumelhart DE, Hilton GE, Williams RJ (1986) Learning representations by back-propagating errors. Nature 323(9):533–536
35. Yu CC, Liu BD (2002) A backpropagation algorithm with adaptive learning rate and momentum coefficient. Nature 323(9):1218–1223
36. Riedmiller M, Braun H (1993) A direct adaptive method for faster backpropagation learning: the RPROP algorithm. In: Proceeding of ISCIS VII, pp 587–591
37. Dai Q, Liu N (2012) Alleviating the problem of local minima in backpropagation through competitive learning. Neurocomputing 94:152–158
38. Yu XH, Chen GA (1997) Efficient backpropagation learning using optimal learning rate and momentum. Neural Netw 10(3):517–527
39. Fang H, Luo J, Wang F (2012) Fast learning in spiking neural networks by learning rate adaptation. Chin J Chem Eng 20(6):1219–1224
40. Igel C, Husken M (2003) Empirical evaluation of the improved RPROP learning algorithms. Neurocomputing 50:105–123

41. Marquardt DW (1963) An algorithm for least - squares estimation of nonlinear parameters. J Soc Ind Appl Math 11(2):431–441
42. Levenberg K (1944) A method for the solution of certain non-linear problems in least squares. Q J Appl Math 2(2):164–168
43. Kermani BG, Schiffman SS, Nagle HT (2005) Performance of the Levenberg Marquardt neural network training method in electronic nose applications. Sensors Actuators B Chem 110(1):13–22
44. Yetilmezsoy K, Demirel S (2008) Artificial neural network (ANN) approach for modeling of Pb(II) adsorption from aqueous solution by Antep pistachio (Pistacia Vera L.) shells. J Hazard Mater 153(3):1288–1300
45. Mukherjee I, Routroy S (2012) Comparing the performance of neural networks developed by using Levenberg Marquardt and Quasi Newton with the gradient descent algorithm for modelling a multiple response grinding process. Expert Syst Appl 39(3):2397–2407
46. Hagan MT, Demuth HB, Bealetitle M (2002) Neural network design. Hagan Publishing, Manchester, pp 12.19–12.27
47. Aydogan B, Ayat B, Ozturk MN, Cevik EO, Yuksel Y (2010) Current velocity forecasting in straits with artificial neural networks, a case study: Strait of Istanbul. Ocean Eng 37(5):443–453
48. Lau HCW, Ho GTS, Zhao Y (2013) A demand forecast model using a combination of surrogate data analysis and optimal neural network approach. Decis Support Syst 54(3):1404–1416
49. Laboissiere LA, Fernandes R, Lage GG (2015) Maximum and minimum stock price forecasting of Brazilian power distribution companies based on artificial neural networks. Appl Soft Comput 35:66–74
50. Mohammadi K, Eslami HR, Dardashti SD (2005) Comparison of regression, ARIMA and ANN models for reservoir inflow forecasting using snowmelt equivalent (a case study of Karaj). J Agric Sci Technol 7:17–30
51. Valipour M, Banihabib ME, Reza SM (2013) Comparison of the ARMA, ARIMA, and the autoregressive artificial neural network models in forecasting the monthly inflow of Dez dam reservoir. J Hydrol 476:433–441
52. De Gooijer J, Hyndman R (2006) 25 years of time series forecasting. Int J Forecast 22(3):443–473
53. Makridakis S, Andersen A, Carbone R, Fildes R, Hibon M, Lewandowski R, Newton J, Parzen E, Winkler R (1982) The accuracy of extrapolation (time series) methods: results of a forecasting competition. J Forecast 1(2):111–153
54. Krause P, Boyle DP, Bäse F (2005) Comparison of different efficiency criteria for hydrological model assessment. Adv Geosci 5(5):89–97
55. Willmott CJ, Robeson SM, Matsuura K (2012) Short communication a refined index of model performance. Int J Climatol 32:2088–2094
56. Shirazi CH, Mosleh A (2010) Data-informed calibration of experts in Bayesian framework. In: Proceeding of the European safety and reliability conference, ESREL 2009, Prague. Taylor & Francis Group, Abingdon, pp 49–55
57. Yafee RA, McGee M (2000) An introduction to time series analysis and forecasting: with applications of SAS and SPSS. Academic Press, Cambridge

Chapter 2
Forecasting Based on Hankel Singular Value Decomposition

2.1 Singular Value Decomposition

Singular Value Decomposition is an old technique that has long been appreciate in the theory of matrices from the early history of SVD of Stewart in 1993 [1].

The SVD is closely related to the spectral decomposition [1]. It was discovered that SVD can be used to derive the polar decomposition of Autonne in which a matrix is factored into the product of a Hermitian matrix and a unitary matrix [2]. SVD was initially applied for square matrices and after with the work of Eckart and Young in 1930, it was extended to rectangular matrices [3].

Principal Component Analysis is a statistical technique for dimensionality reduction which computation is based on the SVD of a positive-semidefinite symmetric matrix. However, it is generally accepted that the earliest descriptions of PCA were given by Pearson (1901) and Hotelling (1933) [4].

Gene Golub published the first effective algorithm in 1965. The algorithm provides essential information about the mathematical background required for the production of numerical software [5, 6].

SVD has been widely used in many fields in recent years. Popular application of SVD were found for denoising [7, 8], features reduction [9, 10], and image compression [11–13].

Important advances have been observed in signals separation. For instance SVD in conjunction with a neuro-fuzzy inference system were used to separate the foetal ECG from the maternal ECG [14].

Zhao and Ye in 2009 demonstrated that a signal can be decomposed into the linear sum of a series of component signals by Hankel matrix-based SVD, and what these component signals reflect in essence are the projections of original signal on the orthonormal bases of M-dimensional and N-dimensional spaces [15]. The SVD

© Springer International Publishing AG, part of Springer Nature 2018
L. M. Barba Maggi, *Multiscale Forecasting Models*,
https://doi.org/10.1007/978-3-319-94992-5_2

application of Zhao and Ye in 2011 was oriented to reduce the noise in a signal caused by gear vibration in headstock [16].

2.1.1 Eigenvalues and Eigenvectors

An eigenvector of a given transformation represents a direction, such that any vector in that direction, when transformed, will continue to point in the same direction, but will be scaled by the eigenvalue [17]. Consider a linear transformation $a:X \rightarrow X$ (the domain is the same as the range). Those vectors $C \in X$ that are not equal to zero and those scalars λ that satisfy:

$$a(v) = \lambda v, \tag{2.1}$$

are called eigenvectors (v) and eigenvalues (λ). The term eigenvector is used here, but v is really a vector space, since v satisfies the equation, then $a * v$ will also satisfy it.

The matrix representation for previous equation can be written as

$$A * V = \lambda * V, \tag{2.2}$$

or

$$[A - \lambda I]V = 0, \tag{2.3}$$

where I is the identity matrix.

This means that the columns of $\|[A - \lambda I]\|$ are dependent, and therefore the determinant of this matrix must be zero:

$$\|[A - \lambda I]\|V = 0, \tag{2.4}$$

This determinant is an jth order polynomial, with j roots.

2.1.2 Matrix Diagonalization

The diagonalization process of a matrix is defined as follows.

Let

$$V = [C_1, C_2, \ldots, C_K], \tag{2.5}$$

where C_i, for $i = 1, \ldots, K$ are the eigenvectors of matrix A. Then,

$$
\left[C^{-1}AC \right] = \begin{bmatrix} \lambda_1 & 0 & \ldots & 0 \\ 0 & \lambda_2 & \ldots & 0 \\ \vdots & \vdots & & \vdots \\ 0 & 0 & \ldots & \lambda_K \end{bmatrix},
\tag{2.6}
$$

where $[\lambda_1, \lambda_2, \ldots, \lambda_K]$ are the eigenvalues of the matrix A.

2.1.3 SVD Theorem

If A is a real $L \times K$ matrix, then there exist orthogonal matrices [6],

$$
U = [U_1, \ldots, U_L] \in \mathbb{R}^{L \times L}, \quad and
\tag{2.7a}
$$

$$
V = [V_1, \ldots, V_K] \in \mathbb{R}^{K \times K},
\tag{2.7b}
$$

such that,

$$
U^T A V = diag(\lambda_1, \ldots, \lambda_L) \in \mathbb{R}^{L \times K}, \quad for \quad L < K,
\tag{2.8}
$$

where both U and V are orthogonal matrices, this means that all their columns are orthonormal. U is orthonormal if $U^T U = I$, therefore the elements $[U_1, \ldots, U_K]$ form an orthonormal basis for $\mathbb{R}^K$, idem for V. The columns of U and V are called eigenvectors of S=A A^T, U is also called EOF (Empirical Orthogonal Function) and left singular vectors of A, whereas V are right singular vectors of A. Scalars λ_i^2 are called eigenvectors of S, while λ_i are singular values of A.

Therefore the SVD expansion of A, is

$$
A = \sum_{i=1}^{L} \lambda_i * U_i * V_i^T,
\tag{2.9}
$$

2.2 Hankel Singular Value Decomposition

The Hankel Singular Value Decomposition technique is used to extract the intrinsic components of low and high frequency from a time series. The process is implemented in three steps: embedding, decomposition, and unembedding [18].

2.2.1 Embedding

A Hankel matrix is used in the first step of the HSVD method. The observed univariate time series $X(n)$ of real values $[x_1, \ldots, x_N]$ is embedded into a matrix $H_{L \times K}$ of Hankel form, which means that all its skew diagonals are constant,

$$
H = \begin{bmatrix} x_1 & x_2 & x_3 & \cdots & x_K \\ x_2 & x_3 & x_4 & \cdots & x_{K+1} \\ \vdots & \vdots & \vdots & \vdots & \vdots \\ x_L & x_{L+1} & x_{L+2} & \cdots & x_N \end{bmatrix}.
$$ (2.10)

Which is equivalent to:

$$
H_1 = \begin{bmatrix} h_{11} & h_{12} & h_{13} & \ldots & h_{1K} \\ h_{21} & h_{22} & h_{23} & \ldots & h_{2K} \\ \vdots & \vdots & \vdots & \vdots & \vdots \\ h_{L1} & h_{L2} & h_{L3} & \ldots & h_{LK} \end{bmatrix}.
$$ (2.11)

L is called *window length* and K is computed with

$$
K = N - L + 1.
$$ (2.12)

The window length L is an integer, $2 \leq L \leq N$. The selection of L is dependent of the time series characteristics and the analysis purpose. There is no a standard process to select L, therefore some alternatives are proposed through empirical data in the Application section.

2.2.2 Decomposition

Let H be an $L \times K$ real matrix, then there exist a $L \times L$ orthogonal matrix U, a $K \times K$ orthogonal matrix V, and a $L \times K$ diagonal matrix Σ with diagonal entries $\lambda_1 \geq \lambda_2 \geq \ldots \geq \lambda_L$, for $L < K$, such that $U^T HV = S$ and $S = HH^T$. Moreover, the numbers $\lambda_1, \lambda_2, \ldots, \lambda_L$ are uniquely determined by H.

$$
H = U * \Sigma * V^T.
$$ (2.13)

U is the matrix of left singular vectors of H and V is the matrix of right singular vectors of H. Besides, the collection (λ_i, U_i, V_i) is the i-th eigentriple of the SVD of H. Elementary matrices $H_1, \ldots, H_L$ of equal dimension $(L \times K)$ are obtained from each eigentriple (λ_i, U_i, V_i),

$$
H_i = \lambda_i * U_i * V_i^T.
$$ (2.14)

2.2.3 Unembedding

The unembedding process is developed to extract the intrinsic components. Each H_i elementary matrix contains each ith component into its first row and last column. Therefore the elements of the component C_i are,

$$C_i = [H_i(1, 1), H_i(1, 2), \ldots, H_i(1, K), H_i(2, K), \ldots, H_i(L, K)], \qquad (2.15)$$

2.2.4 Window Length Selection

The window length is a critical parameter for HSVD decomposition technique. The selection of an effective window length depends on the problem in hand and on preliminarily information about the time series.

The relative energy of the eigenvalues is used to identify the effective window length. The eigenvalue relative energy Λ is computed with the ratio:

$$\Lambda_i = \frac{\lambda_i}{\sum_{j=1}^{L} \lambda_j} \qquad (2.16)$$

where Λ_i represents the energy concentration in the ith eigenvalue. The window length L is an integer with values $2 \leq L \leq N$, and λ_j is the jth singular value.

Commonly the high energy is concentrated in the first singular value. In spite of in this work is also proposed the usage of the peaks of energy, it can be observed through the differentiation of consecutive values of relative energy. Consider,

$$\Delta_j = \Lambda_i - \Lambda_{i+1}, \qquad (2.17)$$

where $i, j = 1, \ldots, N - 1$. A plot is an effective tool to identify the peaks of energy.

2.3 Forecasting Based on HSVD

The forecasting is described in two stages, preprocessing and estimation. The preprocessing stage is based on Hankel Singular Value Decomposition, which is illustrated in Fig. 2.1. The observed time series is decomposed into components of low and high frequency. The component of low frequency C_L is extracted from the first elementary matrix H_1 which is computed with

$$H_1 = \lambda_1 * U_1 * V_1^T. \qquad (2.18)$$

Fig. 2.1 Proposed
forecasting model based on
HSVD

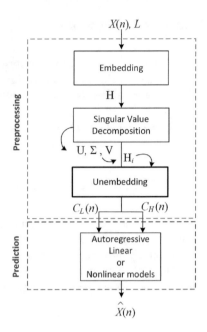

whereas the component of high frequency C_H is computed by simple subtraction as
follows:

$$C_H(n) = X(n) - C_L(n). \tag{2.19}$$

In the prediction stage Autoregressive methods are implemented.

The components $C_L(n)$ and $C_H(n)$ are used to obtain the forecasting $\widehat{X}(n)$. Both
linear and nonlinear autoregressive models are evaluated.

2.3.1   ARIMA Forecasting Based on HSVD

For the linear option, the ARIMA(P, D, Q) model is proposed. Difference is
performed to convert an observed nonstationary time series in stationary. D is
the number of differences. For a stationary time series, the ARMA model is
implemented as follows:

$$\widehat{X}(n) = \sum_{i=1}^{P} \alpha_i Z_i + \sum_{i=1}^{Q} \beta_i \varepsilon_i + \varepsilon(n), \tag{2.20}$$

where P and Q are the orders of the polynomials to the autoregressive part and
moving average part respectively, α and β are the coefficients of the AR terms and

the MA terms respectively, n is the time instant, and Z_i is the ith column of the regressor matrix which is formed with the components $C_L(n)$ and $C_H(n)$ previously extracted,

$$Z = (C_L(n), C_L(n-1), \ldots, C_L(n-P+1), C_H(n), C_H(n-1), \ldots, C_H(n-P+1)]. \quad (2.21)$$

The coefficients α_i, β_i are estimated via Least Square method or Maximum Likelihood Estimation.

2.3.2 ANN Forecasting Based on HSVD

The ANN proposed in this model is a feed-forward sigmoid Multilayer Perceptron (MLP) of three layers [19]. The ANN is denoted with $ANN(P, Q, 1)$, with P inputs, Q hidden nodes, and 1 output. The inputs are the lagged terms contained in each column of the regressor matrix Z (as (2.21)), at hidden layer the sigmoid transfer function is applied, and at output layer the forecasted value is obtained. The forecasting $\widehat{X}$ via ANN is expressed with,

$$\widehat{X}(n+1) = \sum_{j=1}^{Q} b_j Y_{Hj}, \quad (2.22a)$$

$$Y_{Hj} = f\left(\sum_{i=1}^{P} w_{ij} Z_i\right), \quad (2.22b)$$

where $\widehat{X}(n+1)$ is the predicted value at instant $n+1$, b_j and w_{ij} are the linear and nonlinear weights of the ANN connections respectively. $f(.)$ is the sigmoid transfer function denoted with:

$$f(x) = \frac{1}{1+e^{-x}}. \quad (2.23)$$

The parameters b_j and w_{ij} are updated with the application of a learning algorithm.

2.4 Empirical Application: Hankel Singular Value Decomposition for Traffic Accidents Forecasting

2.4.1 Background

Related works about traffic accidents forecasting is scarce, most applications make classification. Classification is presented via Decision rules and trees [20, 21], latent class clustering [22], bayesian networks [23, 24], and genetic algorithm [25]. For traffic accidents forecasting have been implemented Autoregressive Moving Average (ARMA) and ARIMA models [26, 27] and state-space models [28, 29].

The proposed HSVD method is evaluated in combination with ARIMA and an ANN supported by RPROP algorithm for one-step ahead forecasting of traffic accidents. A time series of injured people in traffic accidents coming from Valparaíso, Chile, from year 2003 to 2012 with 531 weekly sampling is used. The performance of HSVD in a linear model and the performance of HSVD in a nonlinear model are evaluated through and HSVD-ARIMA and HSVD-ANN respectively.

2.4.2   Data Description

The empirical data used is the time series of injured people in traffic accidents of Valparaíso, from year 2003 to 2012, they were obtained from CONASET [30]. The data sampling period is weekly, with $N = 531$ registers as it is shown in Fig. 2.2. The data were separated in training sample and testing sample, 85% of samples and 15% respectively.

High oscillation is observed in the original data. The highest peaks are observed at weeks 163, 178, and 365, on January-2006, April-2006, and November-2009 respectively, whereas the lowest peak is observed at weeks 265 and 478 on December/2007 and December/2011 respectively.

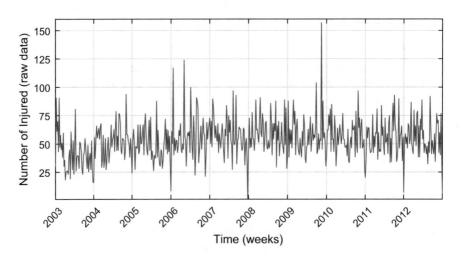

Fig. 2.2 Raw data of injured in traffic accidents, Valparaíso

2.4.3 Data Preprocessing Based on HSVD

HSVD technique is implemented during the preprocessing stage in three steps, embedding, decomposition, and unembedding.

- The original time series X is embedded in a trajectory matrix, then the structure of a Hankel matrix is used to map the data, as (2.10). In this application, the time series is mapped using a Hankel matrix of $L = round(N/2)$ rows. The effective value of L is found by using the observation of the energy peaks of the singular values spectrum. Figure 2.3 shows that the highest quantity of energy is captured by the two first components, therefore in this work has been selected $L = 2$, and consequently two components will be extracted.
- The decomposition process obtains elemental matrices H_1 and H_2, as (2.13) and (2.14).
- The unembedding processes (obtained with (2.15)) extract the components of low frequency C_L and high frequency C_H from H_1 and H_2 respectively.

The first component extracted represents the long-term trend (C_L) of the time series, while the second represents the short-term component of high frequency fluctuation (C_H). The original time series, the component C_L and the component C_H are shown in Fig. 2.4a and b respectively.

2.4.4 Forecasting Models

Two forecasting models are implemented to evaluate the preprocessing through HSVD are:

- Autoregressive linear model: ARIMA, which model is named as HSVD-ARIMA
- Autoregressive nonlinear model: ANN parameterized via RPROP, which model is named as HSVD-ANN

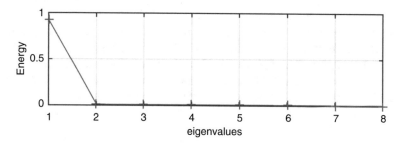

Fig. 2.3 Effective window selection

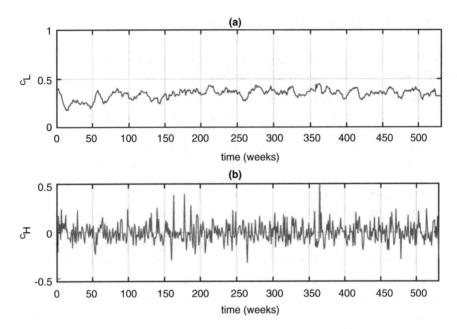

Fig. 2.4 Accidents time series (**a**) Low frequency component, (**b**) High frequency component

2.4.5   HSVD-ARIMA

The P order of the AR model is determined through observation of the Autocorrelation Function (Fig. 2.5). Whereas Q is evaluated using the GCV metric for $1 \leq Q \leq 18$, then the effective value $Q = 11$ is found, as shows Fig. 2.6, therefore the forecasting model is denoted with HSVD-ARIMA(9,0,11).

Once P and Q are found, the forecasting is executed with the testing data set, and the results of HSVD-ARIMA(9,0,11) are shown in Fig. 2.7a, b, and Table 2.1. Figure 2.7a shows the observed values vs. the estimates vales, and a good fit is reached. The relative errors are presented in Fig. 2.7b. The 95% of the points present an error lower than the ±0.5%.

The results presented in Table 2.1 show that high accuracy is achieved with the model HSVD-ARIMA(9,0,11), with a $RMSE$ of 0.00073, a $MAPE$ of 0.26%, a GCV of 0.0013, and 95% of predicted points present an error lower than the ±0.5%. For comparison purposes, Table 2.1 also shows the results that were obtained with the models MA-AR and MA-ANN(LM), which were presented in previous chapter with the same time series.

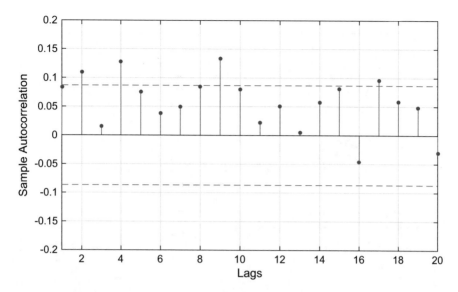

Fig. 2.5 Autocorrelation function of people with injuries in traffic accidents

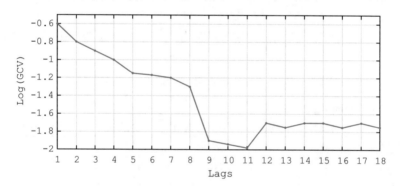

Fig. 2.6 Lags vs GCV to identify the MA order for HSVD-ARIMA

To evaluate the serial correlation of HSVD-ARIMA errors is applied the ACF. Figure 2.8 shows that all the coefficients are inside of the confidence limit, therefore there is no serial correlation. We can conclude that the proposed model HSVD-ARIMA(9,0,11) explains efficiently the variability of the process (Fig. 2.9).

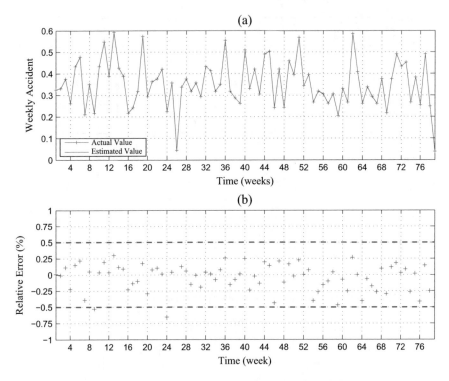

Fig. 2.7 HSVD-ARIMA(9,0,11): (**a**) Observed versus estimated, (**b**) Relative error

Table 2.1 Forecasting results

	HSVD-ARIMA	HSVD-ANN	MA-AR	MA-ANN
$RMSE$	0.00073	0.0038	0.021	0.022
$MAPE$	0.26%	0.76	5.79%	6.76%
GCV	0.0013	0.0049	0.03	0.032
$RE \pm 0.5\%$	95%	–	–	–
$RE \pm 1.5\%$	–	92.3%	–	–
$RE \pm 10\%$	–	–	90.1%	90.7%

2.4.6 HSVD-ANN

The forecasting strategy presented in Fig. 2.1 through the nonlinear autoregressive alternative is detailed here. The HSVD decomposition method was performed to obtain the intrinsic components of low and high frequency. The forecasting model is denoted with $ANN(K, Q, 1)$ which represent $K = 9$ inputs (lagged values of intrinsic components), $Q = 11$ hidden nodes, and 1 output.

The process was ran 30 times, and the first best result was reached at run 21, as Fig. 2.10a shows. The RMSE metric for the best run is presented in Fig. 2.10b.

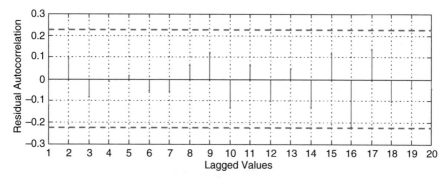

Fig. 2.8 ACF of residuals for HSVD-ARIMA(9,0,11)

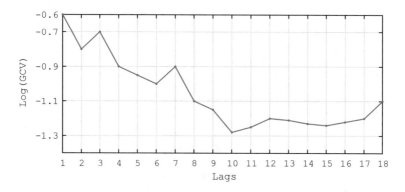

Fig. 2.9 Lags vs GCV to identify the MA order for MA-ARIMA

The evaluation that was executed in the testing stage is presented in Fig. 2.11a, b, and Table 2.1.

Table 2.1, also includes the results that were obtained in Chap. 1 trough MA-ARIMA and MA-ANN. The results show a that high accuracy is obtained with the models based on HSVD.

The serial correlation of the model errors is performed via ACF computation. Figure 2.12 shows that all the coefficients are inside of the confidence limit, and statistically equal to zero, therefore there is no serial correlation. Therefore the proposed model HSVD-ANN(9,11,1) explains efficiently the variability of the process.

2.4.7   Pitman's Correlation Test

The Pitman's correlation test [31] is used to compare the forecasting models in a pairwise fashion. The Pitman's test is equivalent to testing if the correlation (Corr)

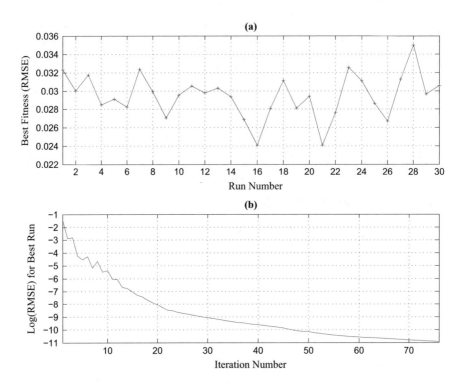

Fig. 2.10 HSVD-ANN(9,11,1) **(a)** Run vs. fitness for 70 iterations and **(b)** Iterations number for the best run

between Υ and Ψ is significantly different from zero, where Υ and Ψ are defined with:

$$\Upsilon = E_1(n) + E_2(n), n = 1, 2, \ldots, N_v \qquad (2.24a)$$

$$\Psi = E_1(n) - E_2(n), n = 1, 2, \ldots, N_v \qquad (2.24b)$$

where E_1 and E_2 represent the one-step-ahead forecast errors for model 1 and model 2 respectively. The null hypothesis is significant at the 5% significance level if $|Corr| > 1.96/\sqrt{N_v}$.

The evaluated correlations between Υ and Ψ are presented in Table 2.2.

The results presented in Table 2.2 show that statistically there is a significant superiority of HSVD-ARIMA model with respect to the rest of models. The results are presented from left to right, being the first the best model and the last the worst model.

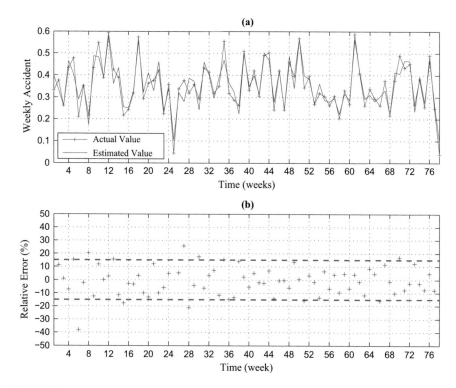

Fig. 2.11 MA-ANN(9,10,1) (**a**) Observed versus estimated and (**b**) Relative error

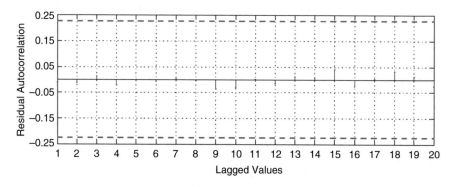

Fig. 2.12 ACF of residuals for HSVD-ANN(9,11,1)

Table 2.2 Pitman's correlation (Corr) for pairwise comparison four models at 5% of significance, and the critical value 0.2219

Models	M1	M2	M3	M4
M1:=HSVD-ARIMA	–	−0.9146	−0.9983	−0.9994
M2:=HSVD-ANN	–	–	−0.9648	−0.9895
M3:=MA-ARIMA	–	–	–	−0.5129
M4:=MA-ANN	–	–	–	–

2.5 Chapter Conclusions

In this chapter was presented the design of a new strategy to decompose a time series of low and high frequency. The method is denominate HSVD due to the Singular Value Decomposition of a Hankel matrix. Two forecasting models were implemented to evaluate the proposed decomposition, one based on Autoregressive Integrated Moving Average (HSVD-ARIMA) and other based on an Artificial Neural Network (HSVD-ANN).

The empirical application shown HSVD-AR and HSVD-ANN in comparison with forecasting based on conventional smoothing by means of moving average combined with ARIMA and ANN (MA-ARIMA and MA-ANN). A weekly sampling of 531 records from 2003:1 to 2012:12 of people with injuries in traffic accidents in Valparaíso were used. Four metrics were used to evaluate the accuracy of all models. RMSE, MAPE, GCV and Relative Error. It was observed that HSVD in conjunction with ARIMA achieved the best accuracy with a $MAPE$ of 0.26%, while the second best was HSVD-ANN, with a $MAPE$ of 0.76%. The Pitman's test was executed to evaluate accuracy differences between the four models. It was found that there is a significant superiority of the forecasting model based on HSVD-ARIMA at 95% confidence level with respect to the regarding models HSVD-ANN, MA-ARIMA and MA-ANN.

References

1. Stewart GW (1993) On the early history of the singular value decomposition. SIAM Rev 35(4):551–566
2. Autonne L (1902) Sur les groupes linéaires, réels et orthogonaux. Bull Soc Math 30:121–134
3. Eckart G, Young G (1936) The approximation of one matrix by another of lower rank. Psychometrika 1(3):211–218
4. Jolliffe IT (2002) Principal component analysis, 2nd edn. Springer, New York
5. Golub GH (1965) Numerical methods for solving linear least squares problems. Numer Math 7:206–216
6. Golub GH, Van Loan CF (1996) Matrix computations. The Johns Hopkins University Press, Baltimore, pp 48–75
7. Abu-Shikhah N, Elkarmi F (2011) Medium-term electric load forecasting using singular value decomposition. Energy 36(7):4259–4271
8. Jha SK, Yadava RDS (2011) Denoising by singular value decomposition and its application to electronic nose data processing. IEEE Sensors J 11(1):34–44

9. Gao J, Kwan PW, Guo Y (2009) Robust multivariate L1 principal component analysis and dimensionality reduction. Neurocomputing 72(4):1242–1249

10. Pozo C, Ruíz-Femenia R, Caballero J, Guillén-Gosálbez G, Jiménez L (2012) On the use of Principal Component Analysis for reducing the number of environmental objectives in multi-objective optimization: application to the design of chemical supply chains. Chem Eng Sci 69(1):146–158

11. Andrews HC, Patterson CL (1976) Singular value decomposition (SVD) image coding. IEEE Trans Commun 24:425–432

12. McGoldrick CS, Bury, A, Dowling WJ (1995) Image coding using the singular value decomposition and vector quantization. In: Fifth international conference on image processing and its applications. IEEE Xplore, Edinburgh, pp 296–300

13. Ranade A, Mahabalarao SS, Kale S (2007) A variation on SVD based image compression. Image Vis Comput 25(6):771–777

14. Al-Zaben A, Al-Smadi A (2005) Extraction of foetal ECG by combination of singular value decomposition and neuro-fuzzy inference system. Phys Med Biol 51(1):137–143

15. Zhao X, Ye B (2009) Similarity of signal processing effect between Hankel matrix-based {SVD} and wavelet transform and its mechanism analysis. Mech Syst Signal Process 23(4):1062–1075

16. Zhao X, Ye B (2011) Selection of effective singular values using difference spectrum and its application to fault diagnosis of headstock. Mech Syst Signal Process 25(5):1617–1631

17. Hagan MT, Demuth HB, Bealetitle M (2002) Neural network design. Hagan Publishing, Manchester, pp 12.19–12.27

18. Barba L, Rodríguez N, Montt C (2014) Smoothing strategies combined with ARIMA and neural networks to improve the forecasting of traffic accidents. Sci World J 2014:12 pp, Article ID 152375

19. Freeman JA, Skapura DM (1991) Neural networks, algorithms, applications, and programming techniques. Addison-Wesley, Boston

20. Joaquín Abellán J, López G, de Oña J (2013) Analysis of traffic accident severity using Decision Rules via Decision Trees. Expert Syst Appl 40(15):6047–6054

21. Chang LY, Chien JT (2013) Analysis of driver injury severity in truck-involved accidents using a non-parametric classification tree model. Saf Sci 51(1):17–22

22. de Oña J, López G, Mujalli R, Calvo FJ (2013) Analysis of traffic accidents on rural highways using Latent Class Clustering and Bayesian Networks. Accid Anal Prev 51:1–10

23. Deublein M, Schubert M, Adey BT, Köhler J, Faber MH (2013) Prediction of road accidents: a Bayesian hierarchical approach. Accid Anal Prev 51:274–291

24. Mujalli RO, López G, Garach L (2016) Bayes classifiers for imbalanced traffic accidents datasets. Accid Anal Prev 88:37–51

25. Manuel Fogue M, Garrido P, Martinez FJ, Cano JC, Calafate CT, Manzoni P (2013) A novel approach for traffic accidents sanitary resource allocation based on multi-objective genetic algorithms. Expert Syst Appl 40(1):323–336

26. Quddus MS (2008) Time series count data models: an empirical application to traffic accidents. Accid Anal Prev 40(5):1732–1741

27. García-Ferrer A, de Juan A, Poncela P (2006) Forecasting traffic accidents using disaggregated data. Int J Forecast 22:203–222

28. Commandeur JF, Bijleveld FD, Bergel-Hayat R, Antoniou C, Yannis G, Papadimitriou E (2013) On statistical inference in time series analysis of the evolution of road safety. Accid Anal Prev 60:424–434

29. Antoniou C, Yannis G, (2013) State-space based analysis and forecasting of macroscopic road safety trends in Greece. Accid Anal Prev 60:268–276

30. CONASET (2015) Comisión Nacional de Seguridad de Tránsito. http://www.conaset.cl

31. Hipel K, McLeod A (1994) Time series modelling of water resources and environmental systems. Elsevier, New York

Chapter 3
Multi-Step Ahead Forecasting

3.1 Background

The accumulation of errors and lack of information make multi-step-ahead forecasting more difficult [1]. It has been probed that such difficulties increase in nonlinear time series [2, 3]. Besides, multi-step ahead prediction overhead the one-term delay that could suffer one-step ahead prediction through nonlinear models as learning machines. For instance when the current value of the signal is very close to the next consecutive term, the learning machine will provide a local minimum on error surface, which may be a suboptimal solution for an optimization problem [4].

Iterative and Direct are multi-step strategies that have been widely used along the literature review [5]. The iterative method implement h times a one-step ahead forecasting model to obtain the h forecasts, the output of each repetition is fed back as an input for the next repetition. In contrast the direct strategy estimates a set of h forecasting models, each one returns the forecast for an specific ith horizon in the range $(1, \ldots, h)$.

In the iterative method, the use of the predicted values as inputs deteriorates the accuracy of the prediction, whereas the direct strategy increases the complexity of the prediction, but more accurate results are achieved [6]. The superiority of the direct method against the iterative method has been pointed out in diverse type of data such as, chaotic time series [7], multi-periodic of different nature [8], environmental [9], energy [10], among others.

Hankel Singular Value Decomposition is a new preprocessing technique that was shown in previous chapter to extract intrinsic components of low and high frequency from a time series. Singular Spectrum Analysis (SSA) is a nonparametric technique of time series analysis which incorporates SVD for filtering time series that will be explained in next sections.

© Springer International Publishing AG, part of Springer Nature 2018
L. M. Barba Maggi, *Multiscale Forecasting Models*,
https://doi.org/10.1007/978-3-319-94992-5_3

3.2   Strategies for Multi-Step Ahead Forecasting

The conventional strategies for multi-step ahead forecasting are detailed in this section, iterative, direct and MIMO.

3.2.1   Iterative Strategy

The iterative strategy is also called recursive or multi-stage strategy. This strategy is implemented through h repetitions to obtain the h forecasts. The output of each repetition is fed back as an input for the next repetition.

Consider an univariate time series x of length N, the one-step ahead prediction model depends on the past $P < N$ points, and it has the form:

$$\widehat{X}(n+1) = f\,[X(n), X(n-1), \ldots, X(n-P+1)]\,, \tag{3.1}$$

where n is the time instant and P is the length of the regressor vector (also called time window size).

Now consider multi-step ahead prediction of the same time series, specifically τ-step ahead prediction. The iterative strategy computes τ one-step ahead predictions.

Mathematically τ-step ahead prediction is expressed as

$$\widehat{X}(n+h) = f\,[X(n+h-1), X(n+h-2), \ldots, X(n+h-P)]\,, \tag{3.2}$$

where $h = 1, \ldots, \tau$,

The iterative strategy is based on τ iterations of the exact one-step model to τ-step ahead predictions.

The usage of the past predicted values as input of new predictions presents the error of accumulation problem. The single-step predictor must be accurate enough that the accumulation error in each iteration be as small as possible, however it is not possible trying with nonlinear and/or noisy signals. The prediction performance deteriorates significantly in the presence of noise and outliers [7].

3.2.2   Direct Forecasting Strategy

The direct strategy estimates τ models between regressors to compute τ-step ahead prediction. Each model returns the direct forecast of $\widehat{X}$. This strategy can be expressed as

$$\widehat{X}(n+h) = f\,[Z(n), Z(n-1), \ldots, Z(n-P+1)]\,, \tag{3.3}$$

where n is the time instant, $h = 1, \ldots, \tau$, where τ is the maximum horizon prediction. Z_i is the ith input regressor matrix, and P is the regression order, also called time window size.

The input regressor matrix Z contains the lagged values of C_L and C_H, each one of size $N_t \times P$, where N_t is the number of samples for training and testing.

The matrix for of AR via Direct Strategy is defined with:

$$\hat{X} = \beta Z^T, \tag{3.4}$$

where β is the vector of linear coefficients with $2P$ dimension, and Z^T is the transposed regressor matrix. Z has order N_t, where N_t is the number of samples. The matrix of coefficients β is computed with the Least Square Method (LSM) as below:

$$\beta = XZ^\dagger, \tag{3.5}$$

where X is a $N_t \times 2P$ matrix of observed samples. $Z^\dagger$ is the Moore-Penrose pseudoinverse matrix of Z.

3.2.3 MIMO Forecasting Strategy

MIMO computes the output for the forecast horizon in a single simulation with an unique model. The output of MIMO is a matrix rather than a vector [11]. MIMO strategy is defined as follows:

$$\left[\widehat{X}(n + 1), \widehat{X}(n + 2), \ldots, \widehat{X}(n + \tau)\right] = f\left[Z(n), Z(n - 1), \ldots, Z(n - P + 1)\right]. \tag{3.6}$$

where τ is the forecast horizon, $\widehat{X}$ is the matrix of outputs. Each column will contain the prediction related to a specific horizon.

The following equation defines MIMO in matrix form:

$$\hat{X} = \beta Z^T, \tag{3.7}$$

where β is a $\tau \times 2P$ matrix of linear coefficients and Z^T is the autoregressive transposed matrix. Z is $N_t \times 2P$ dimension, where N_t is the number of samples. The matrix of coefficients β is computed with the Least Square Method (LSM) as below:

$$\beta = XZ^\dagger \tag{3.8}$$

where $Z^\dagger$ is the Moore-Penrose pseudoinverse matrix of Z.

3.3   Singular Spectrum Analysis

The Singular Spectrum Analysis (SSA) method was developed by Broomhead and King in 1986 for extraction of qualitative information from dynamic systems [12].

SSA has been applied in numerous fields to decompose a time series in a sum of components of different nature, as trend, cyclic component, periodic component, and an irregular component. The implementation of SSA is conventionally performed in four steps: embedding, decomposition, grouping, and diagonal averaging [13]. Most applications based on SSA has been conventionally used to separate signal and noise [14–16].

In this section SSA is applied to extract components of low and high frequency from a nonstationary time series.

3.3.1   Embedding

A data points sequence observed $X(n)$ is embedded in a trajectory matrix H, called Hankel matrix. The embedding space $\mathbb{R}^N$, is the space of all N-element and is the natural object of interest.

The embedding uses a sequence of L multidimensional lagged vectors of length K, or a sequence of K multidimensional lagged vectors of length L,

$$H_{L \times K} = \begin{pmatrix} x_1 & x_2 & x_3 & \dots & x_K \\ x_2 & x_3 & x_4 & \dots & x_{K+1} \\ \vdots & \vdots & \vdots & \vdots & \vdots \\ x_L & x_{L+1} & x_{L+2} & \dots & x_N \end{pmatrix}, \tag{3.9}$$

where the elements $H(i, j) = x_{i+j-1}$, L is the window length with values $2 \leq L \leq N/2$, while K is computed as follows

$$K = N - L + 1. \tag{3.10}$$

In comparison with HSVD, SSA implements the same process to map the observed time series into a Hankel matrix.

3.3.2   Decomposition

The Singular Value Decomposition of the trajectory matrix H has the form

$$H = \sum_{i=1}^{L} \sqrt{\lambda_i} U_i V_i^\top, \tag{3.11}$$

where each λ_i is the ith eigenvalue of the matrix $S = HH^T$ arranged in decreasing order of magnitudes. The orthonormal eigenvectors system of H is $U_1, \ldots, U_L$.

Standard SVD terminology calls $\sqrt{\lambda_i}$ the ith singular value of the matrix H. The left singular vectors are U_i and V_i are the right singular vectors of H.

The SVD of the trajectory matrix H can also be written as

$$H = H_1 + \ldots + H_d, \tag{3.12}$$

where $d = maxi$, such that $\lambda_i > 0$.

The matrices H_i have rank 1, therefore they are *elementary matrices*. The collection $[\sqrt{\lambda_i} U_i V_i^T]$ is called ith eigentriple of the SVD.

The decomposition step is also equivalent for both HSVD and SSA methods.

3.3.3 Grouping

From the elementary matrices, the grouping step is executed to create partitions of the set of indices $1, \ldots, d$ into m disjoint subsets $I_1, \ldots, I_m$.

For instance, let $I_1 = (i_1, \ldots, i_p)$, the *resultant matrix* H_{I1} is obtained with,

$$H_{I1} = H_{i_1} + H_{i_2} + \ldots + H_{i_p}, \tag{3.13}$$

The process is repeated until to compute the H_{Im} grouped matrix.

3.3.4 Diagonal Averaging

The step of Diagonal Averaging transforms each matrix of the grouped decomposition into new series of equal length as the observed time series.

Let H_{Ij}, with $j = 1, \ldots, m$ be an $L \times K$ matrix with elements h_{ij}, $1 \leq i \leq L$, $1 \leq j \leq K$. The settings $L* = min(L, K)$, $K* = max(L, K)$, and $N = L + K - 1$ are performed. Let $h_{ij}^* = h_{ij}$ if $L < K$ and $h_{ij}^* = h_{ji}$ otherwise.

Assuming that $L < K$, diagonal averaging transfers the matrix H_{Ij} of $L \times K$ dimension to the series $C_1, \ldots, C_m$. For instance:

$$C_{1k} = \begin{cases} \frac{1}{k-1} \sum_{l=1}^{k} H(l, k - l), & 2 \leq k \leq L, \\ \frac{1}{L} \sum_{l=1}^{L} H(l, k - l), & L < k \leq K + 1, \\ \frac{1}{K+L-k+1} \sum_{l=k-K}^{L} H(l, k - l), & K + 2 \leq k \leq K + L. \end{cases} \tag{3.14}$$

3.4   Windows Length Selection

The time series of length p is embedded into a trajectory matrix of dimension $r \times q$, where r is the window length. The general rule used to delimit the windows length is $2 \le r \le p/2$; a large decomposition is given with a high value of r, while a short decomposition is the opposite. During the literature review, some strategies have been used to select the window length; some instances are $r = p/4$ [17], weighted correlation [18] and extreme of autocorrelation [19].

The method used in this chapter to select the effective window length is based on the relative energy of the singular values. A detailed description will be presented in Empirical Application Section.

3.5   Forecasting Based on Components of Low and High Frequency

The forecasting methodology presented here is composed of two stages, preprocessing and prediction, a block diagram is shown in Fig. 3.1. The preprocessing stage extracts the components of low and high frequency from an observed time series via HSVD or SSA and the prediction is developed using an Autoregressive (AR) model or an ANN.

Fig. 3.1 Block diagram of forecasting methodology

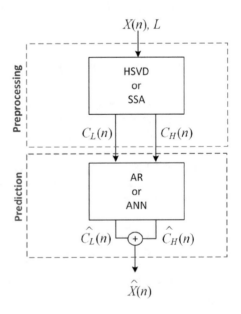

3.5.1 Components Extraction Based on HSVD and SSA

Both HSVD and SSA extract the component of low frequency from the first elementary matrix H_1, whereas the component of high frequency C_H is computed by subtraction between the observed signal and C_L,

$$C_H(n) = X(n) - C_L(n). \tag{3.15}$$

In the prediction stage, the components $C_L(n)$ and $C_H(n)$ are predicted separately. Their addition is used to obtain the prediction $\widehat{X}(n)$,

$$\widehat{X}(n+1) = \widehat{c_L}(n+1) + \widehat{c_H}(n+1). \tag{3.16}$$

3.5.2 Multi-Step Ahead Forecasting via AR Model

The prediction of C_L is defined with

$$\widehat{C}_L(n+1) = \sum_{i=1}^{P} \alpha_i Z_{Li}, \tag{3.17a}$$

$$Z_L = (C_L(n), C_L(n-1), \ldots, C_L(n-P+1)). \tag{3.17b}$$

where P is the order of the autoregressive model (number of lagged values) for C_L prediction and $alpha_i$ is the ith linear coefficient for $i = 1, \ldots, P$.

Whereas the $\widehat{C}_H$ is computed with both components C_H and C_L. The component C_L is used as external input:

$$\widehat{C}_H(n+1) = \sum_{i=1}^{2P} \beta_i Z_{Hi}, \tag{3.18a}$$

$$Z_H = (C_H(n), C_H(n-1), \ldots, C_H(n-P+1), C_L(n),$$
$$C_L(n-1), \ldots, C_L(n-P+1),). \tag{3.18b}$$

where β_i is the ith coefficient for $i = 1, \ldots, 2P$.

The coefficients estimation is based on Least Square method. The components C_L and C_H are represented with the linear relationship between the coefficients and the regressor matrices Z_L and Z_H. This is expressed in matrix form as follows:

$$C_L = \alpha Z_L, \tag{3.19a}$$

$$C_H = \beta Z_H, \tag{3.19b}$$

where Z_L is the regressor matrix of N rows and P columns (in this case N is the sample size) and α a vector of P rows and one column; while Z_H is the regressor matrix of N rows and $2P$ columns, and β is a vector of $2P$ rows and one column. The coefficients are computed by using the Moore-Penrose pseudoinverse $Z_L\dagger$, and $Z_H\dagger$ as follows

$$\alpha = Z_L\dagger C_L, \tag{3.20a}$$

$$\beta = Z_H\dagger C_H. \tag{3.20b}$$

3.5.3   Multi-Step Ahead Forecasting via ANN

A conventional Artificial Neural Network is described for both hybrid models: HSVD-ANN and SSA-ANN. A sigmoid Multilayer Perceptron (MLP) of three layers [20], which is denoted with $ANN(P, Q, 1)$. The inputs are the P lagged terms contained in the regressor matrix Z. The hidden layer has Q nodes and the output has one node.

The prediction $\widehat{C}_L(n + 1)$ via ANN is expressed with,

$$\widehat{C}_L(n + 1) = \sum_{j=1}^{Q} b_j Y_{Hj}, \tag{3.21a}$$

$$Y_{Hj} = f\left(\sum_{i=1}^{P} w_{ij} Z_{Li}\right), \tag{3.21b}$$

where Y_{Hj} is the jth hidden level output. b_j is the jth linear weight of the connection between the jth hidden node and the output. w_{ij} is the nonlinear weight of the connection between the ith input node and the jth hidden node. Z_{Li} represents the i-th lagged vector of low frequency.

The prediction of $\widehat{C}_H(n + 1)$ via ANN is expressed with,

$$\widehat{C}_H(n + 1) = \sum_{j=1}^{Q} b_j Y_{Hj}, \tag{3.22a}$$

$$Y_{Hj} = f\left(\sum_{i=1}^{P} w_{ij} Z_{Hi}\right), \tag{3.22b}$$

where Z_{Hi} represents the i-th lagged vector of high frequency.

The sigmoid transfer function is denoted with:

$$f(x) = \frac{1}{1 + e^{-x}}. \tag{3.23}$$

Parameters b and w are updated with the application of the learning algorithm.

3.6   Empirical Application: Multi-Step Ahead Forecasting of Traffic Accidents Based on HSVD and SSA via Direct Strategy

3.6.1   Background

The data provided by the Chilean Commission of Traffic Safety (CONASET) [21], shows a high and increasing rate of fatalities and injuries in traffic accidents from 2000 through 2014. Santiago is the most populated Chilean region, which time series of injured people is used in this analysis. The abundance of information and research about traffic accidents, severities, risks, occurrence factors, among others, might appear somewhat unapproachable in terms of the prevention plans.

In this case study, with categorical causes defined by CONASET and the ranking method, the potential causes of injured in traffic accidents, which can be counteracted through campaigns directed to the change of attitude in drivers, passengers, and pedestrians toward road safety.

Two groups have been created based on the primary and secondary causes which are present in around 75% of injured, and a third group was created with the remaining 25%. At first, nonstationary and nonlinear characteristics were found in the time series, turning the forecasting in a difficult task.

In this work, the SSA implementation is simplified in three steps: embedding, decomposition, and diagonal averaging, only one elementary matrix is needed and it is computed with the first SVD eigentriple. The main difference observed in the implementation process of HSVD and SSA is the step performed for components extraction. HSVD avoids diagonal averaging, instead unembedding is proposed.

3.6.2   Models in Evaluation

The forecasting methodology applied in the analyzed hybrid models is described in two stages, preprocessing and prediction, as Fig. 3.1 illustrates. In the preprocessing stage Singular Spectrum Analysis and Singular Value Decomposition of a Hankel matrix are used to extract an additive component of low frequency from the observed time series, and by simple subtraction between the observed time series and the component of low frequency, the component of high frequency is obtained. In the prediction stage, linear and nonlinear models are implemented.

3.6.3 Data Description

The Chilean Police (Carabineros de Chile) collects the features of the traffic accidents, and CONASET records the data. The regional population is estimated in 6,061,185 inhabitants, equivalent to 40.1% of the national population. Santiago shows a high rate of the events with severe injuries from 2000 through 2014, with 260,074 injured people.

The entire series of 783 registers is shown in Fig. 3.2, which have been collected in the mentioned period from January to December, with weekly sampling. The highest number of injured was observed in weeks 84, 184, 254, and 280; while the lowest number of injured was observed in weeks 344, 365, 426, 500, and 756.

One hundred causes of traffic accidents have been defined by CONASET and grouped in categories. Initially, the ranking technique was applied to find the potential causes related to inappropriate behavior of drivers, passengers and pedestrians. It was found that 75% of incidents with injured people were due to inappropriate behavior described in 20 causes. Two analysis groups were created from those causes, groups 1 and 2 belongs to categories: *reckless driving, recklessness in passenger, recklessness in pedestrian, alcohol in driver, alcohol in pedestrian*, and *disobedience to signal*. Injured-G1 involves 60% with 10 principal causes, whereas Injured-G2 involves the regarding 15% with the 10 secondary causes. Table 3.1 shown the details of causes for both group 1 and 2.

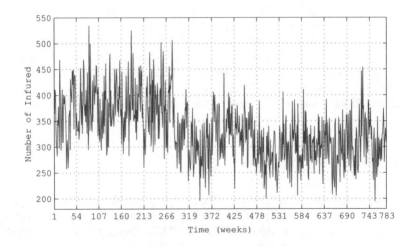

Fig. 3.2 Injured people in traffic accidents due to 100 causes

Table 3.1 Causes of injured in traffic accidents, group 1 and group 2

Category	Num	Cause	Importance
Imprudent driving	1	Unwise distance	1
	2	Inattention to traffic conditions	2
	3	Disrespect to pedestrian passing	8
	4	Disrespect for give the right of way	9
	5	Unexpected change of track	10
	6	Improper turns	11
	7	Overtaking without enough time or space	14
	8	Opposite direction	18
	9	Backward driving	19
Disobedience to signal	10	Disrespect to red light	3
	11	Disrespect to stop sign	4
	12	Disrespect to give way sign	6
	13	Improper speed	13
Alcohol in driver	14	Drunk driver	7
	15	Driving under the influence of alcohol	15
Recklessness in pedestrian	16	Pedestrian crossing the road suddenly	5
	17	Reckless pedestrian	12
	18	Pedestrian outside the allowed crossing	17
Recklessness in passenger	19	Get in or get out of a moving vehicle	16
Alcohol in pedestrian	20	Drunk pedestrian	20

From Table 3.1, the causes are listed in order of relevancy. The cause with highest relevance has importance value 1 (with the highest number of injured), and the cause with minor relevance has value 20. It was observed that the first three causes of injured in traffic accidents are: *Unwise Distance, inattention to traffic conditions,* and *disrespect to red light.* The cause with the lowest incidence for injured in traffic accidents is *Drunk pedestrian.* It was also observed that the categories with the highest number of causes are: *imprudent driving* and *disobedience to signal.* Figures 3.3a, 3.4a, and 3.5a show the observed time series of the groups Injured-G1, Injured-G2, and Injured-G3, respectively. The values have been normalized.

The third group labeled with Injured-G3 show injured related to categories: *road deficiencies, mechanical failures,* and *undetermined/non categorized causes.* This group involves around 25% of injured in traffic accidents during the observed period.

3.6.4 Forecasting

Complementary information was observed about traffic accidents conditions with high rate of injured people. About vehicles, automobiles are involved in 54%

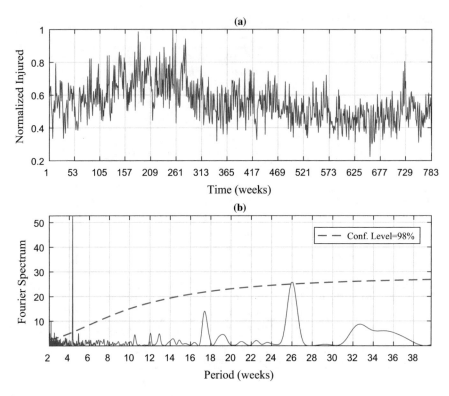

Fig. 3.3 (a) Injured-G1, (b) FPS of Injured-G1. The thick solid line is the global wavelet spectrum for Injured-G1, while the dashed line is the red-noise spectrum

of sinister, followed by vans and trucks with 19%, bus and trolley with 16%, motorcycles and bicycles with 8%, and others with 3%. About environmental conditions, 85% of sinister was observed with cloudless conditions. Additionally 97% of the traffic accidents with injured people have taken place in urban areas, whereas 3% corresponds to rural area. About relative position, 46% of the events have been produced in intersections controlled by traffic signals or police officers with 46%, followed by accidents that happened in straight sections with 37%, and other relative positions with 17%.

Fourier Power Spectrum (FPS) was performed for identifying of relevant periods in the analysis groups. Figures 3.6b, 3.7b, and 3.8b show the FPS of Injured-G1, Injured-G2, and Injured-G3 respectively. From figures, the time series Injured-G1 has similar dynamic with respect to full series, into consideration that this group contains the predominant causes. High data variability is observed during the analyzed period; for instance Injured-G1 presents upward trend between weeks 1 and 280 which is followed by downward trend until week 348, that behavior continues in the remaining observed period. On the other hand, Injured-G2 presents downward trend from week 232 until the end, and Injured-G3 presents upward trend

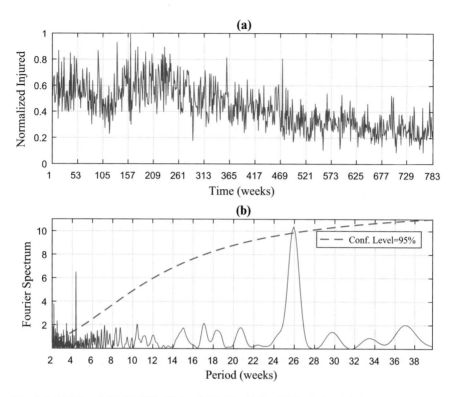

Fig. 3.4 (**a**) Injured-G2, (**b**) FPS of Injured-G2. The thick solid line is the global wavelet spectrum for Injured-G2, while the dashed line is the red-noise spectrum

from week 491 until end. The FPS analysis shows the signal spectrum and the red noise spectrum, a signal spectrum peak is significative when its value is higher than the red noise spectrum [22]. Both series, Injured-G1 and Injured-G2 present the highest peak at week 26 at 98% of confidence level, whereas for Injured-G3 shows the highest peak at week 17 at 73% of confidence level. Whit the peaks information is set the AR order, $P = 26$ for Injured-G1 and Injured-G2, and $P = 17$ for Injured-G3.

Additionally the nonstationary trend in all analysis groups was verified by means of KPSS test [23]. The KPSS test assesses the null hypothesis that the signals are trend stationary, it was rejected at 5% significance level, consequently nonstationary unit-root processes are present in an analyzed signal.

3.6.4.1 Components Extraction

Both preprocessing techniques HSVD and SSA embed the time series in a trajectory matrix. The initial window length used is $L = N/2$, based on the N length of the

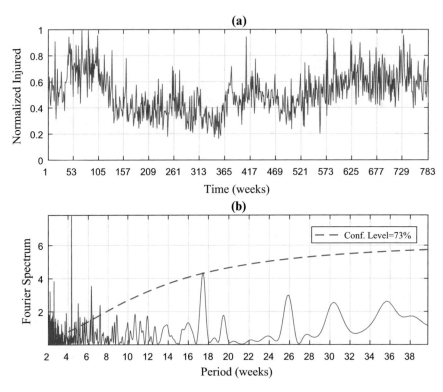

Fig. 3.5 (a) Injured-G3, (b) FPS of Injured-G3. The thick solid line is the global wavelet spectrum for Injured-G3, while the dashed line is the red-noise spectrum

series. The embedding matrix H is then decomposed in singular values and singular vectors.

In this case, a high energy content was observed in the first $t = 20$ eigenvalues. The lowest peaks of differential energy was used and evaluated to set the effective window length. Some experiments were developed examining diverse effective window length values, finally it was chosen by trial and error from those energy peaks that were given by the differential energy of the singular values. Then the effective window length was set in $L = 15$, $L = 17$, and $L = 15$, for Injured-G1, Injured-G2, and Injured-G3 respectively. The embedding process is implemented again with the effective window L, and the decomposition is recomputed following the methodology.

The first elementary matrix H_1 is used to obtain the low frequency C_L component. In HSVD direct extraction (unembedding) is performed from the first row and last column of H_1. While in SSA, diagonal averaging is computed over H_1. Finally the component of high frequency is computed by simple difference.

The components extracted by HSVD and SSA are shown in Figs. 3.9, 3.10 and 3.11. The C_L components extracted by HSVD and SSA from all signals

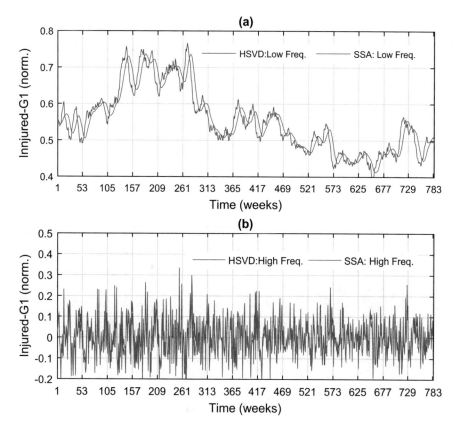

Fig. 3.6 Injured-G1 components (**a**) Low frequency components via HSVD and SSA (**b**) High frequency components via HSVD and SSA

are shown in Figs. 3.6a, 3.7a, and 3.8a. Long-memory periodicity features were observed in all C_L components. The resultant components of high frequency are shown in Figs. 3.6b, 3.7b, and 3.8b, short-term periodic fluctuations were identified in all C_H components.

Injured-G1 signals show that the principal 10 causes present the highest incidence between years 2002 and 2005 (weeks 106–312), it descends from 2006 until half 2012 (around week 710), an increment is observed between weeks 711 and 732 (second semester of 2013 and first semester of 2014).

Injured-G2 signals shown that secondary 10 causes present the highest incidence in years 2000, 2003 and 2004, it descends from 2005, forward downtrend is observed.

Injured-G3 signals shown other causes have the highest incidence in year 2001, whereas from year 2002 presents strong decay until 2006, forward, uptrend is observed with a temporal decreasing in 2009.

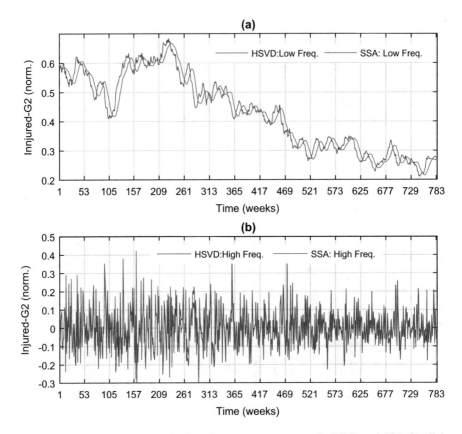

Fig. 3.7 Injured-G2 components (**a**) Low frequency components via HSVD and SSA (**b**) High frequency components via HSVD and SSA

Prevention plans and punitive laws have been implemented in Chile during the analyzed period, vial education, drivers licensing reforms, zero tolerance law, Emilia's law, transit law reforms, among others. The effect of a particular preventive or punitive action is not analyzed in this application, however the proposed short-term prediction methodology based on observed causes and intrinsic components is a contribution to government and society in preventive plans formulation, its implementation and the consequent evaluation.

3.6.4.2   Prediction Results

Each data set of low and high frequency that was obtained in the preprocessing stage has been divided into two subsets, training and testing for prediction. The training subset involves 70% of the samples, and consequently the testing subset involves the remaining 30%.

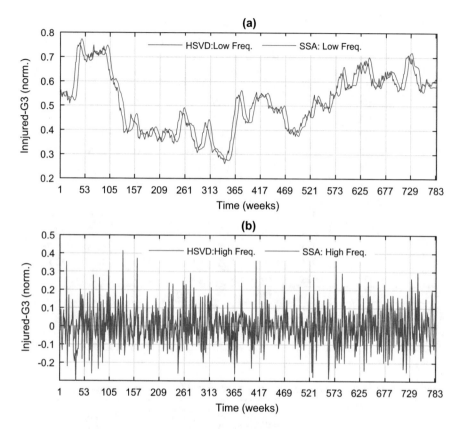

Fig. 3.8 Injured-G3 components (**a**) Low frequency components via HSVD and SSA (**b**) High frequency components via HSVD and SSA

The prediction is implemented by means of the autoregressive models, linear (AR), or nonlinear (ANN). The models use the lagged terms of the components C_L and C_H. The order of the autoregressive model was previously set through the FPS information.

Multi-step ahead forecasting was implemented by means of multiple models in reason that direct strategy is used. The forecasting accuracy is measured with normalized Root Mean Square Error ($nRMSE$) and modified Nash-Sutcliffe Efficiency ($mNSE$). The results are presented in Tables 3.2, 3.3 and 3.4 for Injured-G1, Injured-G2 and Injured-G3 respectively.

Table 3.2, Figs. 3.9 and 3.10, present the prediction results of multi-week ahead prediction of Injured-G1 via HSVD-AR, SSA-AR, HSVD-ANN and SSA-ANN. The results shows that the accuracy decreases as the time horizon increases. The best accuracy was reached by using SSA-AR for 1 to 10-week ahead forecasting while for 11 to 14-week ahead forecasting HSVD-AR is more accurate. SSA-AR present an average $nRMSE$ of 0.013 and an average $mNSE$ of 92.7%, it is followed by

Table 3.2 Multi-step ahead forecasting results for Injured-G1 via direct strategy

h (week)	nRMSE				mNSE (%)			
	HSVD-AR	SSA-AR	HSVD-ANN	SSA-ANN	HSVD-AR	SSA-AR	HSVD-ANN	SSA-ANN
1	0.008	0.001	0.01	0.002	95.7	99.3	93.7	99.1
2	0.011	0.003	0.02	0.003	93.6	98.4	89.4	97.8
3	0.015	0.005	0.03	0.007	91.7	97.4	83.0	96.5
4	0.018	0.007	0.04	0.011	90.1	96.3	79.0	93.8
5	0.020	0.009	0.04	0.014	89.0	95.0	78.8	91.6
6	0.021	0.011	0.04	0.018	88.1	93.6	75.0	88.4
7	0.023	0.014	0.06	0.027	87.0	92.1	67.0	87.1
8	0.024	0.017	0.06	0.036	86.8	90.7	64.4	79.3
9	0.024	0.019	0.07	0.041	86.9	89.3	62.0	75.6
10	0.024	0.022	0.07	0.056	86.8	87.9	61.9	73.5
11	0.023	0.024	0.07	0.056	87.1	86.7	61.5	70.7
12	0.022	0.026	0.07	0.060	87.7	85.6	56.5	67.8
13	0.020	0.028	0.07	–	89.1	84.9	57.7	–
14	0.016	–	0.06	–	91.0	–	–	–
Min	0.008	0.001	0.011	0.002	86.8	84.9	56.5	67.8
Max	0.024	0.028	0.074	0.06	95.7	99.3	93.7	99.1
Mean 1–13	0.19	0.013	0.047	0.028	89.2	92.7	72.7	85.1

Table 3.3 Multi-step ahead forecasting results for Injured-G2 via direct strategy

h (week)	nRMSE				mNSE (%)			
	HSVD-AR	SSA-AR	HSVD-ANN	SSA-ANN	HSVD-AR	SSA-AR	HSVD-ANN	SSA-ANN
1	0.011	0.002	0.01	0.002	96.3	99.4	95.4	99.3
2	0.015	0.004	0.02	0.005	95.3	98.5	93.0	98.5
3	0.018	0.007	0.02	0.008	93.9	97.6	92.1	97.5
4	0.021	0.010	0.03	0.011	92.9	96.6	90.5	96.4
5	0.023	0.013	0.04	0.014	92.2	95.6	89.5	95.2
6	0.025	0.017	0.04	0.018	91.4	94.4	88.3	94.0
7	0.027	0.020	0.04	0.022	90.9	93.3	85.3	92.7
8	0.029	0.024	0.04	0.024	90.3	92.2	84.6	91.6
9	0.030	0.027	0.04	0.032	89.8	91.1	84.1	89.6
10	0.030	0.030	0.05	0.032	89.8	90.1	84.0	88.8
11	0.030	0.033	0.05	0.039	90.1	89.2	82.3	86.9
12	0.028	0.035	0.06	–	90.5	90.5	82.1	–
13	0.027	0.037	0.06	–	90.8	90.8	81.0	–
14	0.026	–	0.07	–	91.2	–	75.4	–
Min	0.011	0.002	0.014	0.002	89.8	87.7	75.4	86.9
Max	0.030	0.037	0.072	0.039	96.3	99.4	95.4	99.3
Mean 1–12	0.023	0.017	0.035	0.019	92.1	94.4	88.1	93.7

Table 3.4 Multi-step ahead forecasting results for Injured-G3 via direct strategy

h (week)	nRMSE				mNSE (%)			
	HSVD-AR	SSA-AR	HSVD-ANN	SSA-ANN	HSVD-AR	SSA-AR	HSVD-ANN	SSA-ANN
1	0.01	0.003	0.01	0.004	95.9	98.3	95.8	98.3
2	0.01	0.007	0.01	0.008	93.4	96.4	93.5	96.0
3	0.02	0.012	0.02	0.0126	91.1	94.1	90.9	94.3
4	0.02	0.017	0.02	0.018	89.2	91.6	87.7	91.4
5	0.02	0.022	0.03	0.023	88.1	89.2	87.1	88.9
6	0.02	0.027	0.03	0.029	87.0	86.7	85.9	86.0
7	0.03	0.032	0.04	0.037	86.1	84.6	82.4	82.5
8	0.03	0.036	0.03	0.181	84.9	82.7	83.0	–
9	0.03	–	0.04	0.208	83.9	–	79.8	–
10	0.03	–	0.04	–	83.1	–	80.8	–
11	0.03	–	0.06	–	82.9	–	71.3	–
12	0.03	–	0.06	–	82.9	–	69.1	–
13	0.03	–	0.05	–	83.9	–	77.0	–
14	0.03	–	–	–	85.3	–	–	–
Min	0.010	0.003	0.009	0.004	82.9	82.7	77.0	82.5
Max	0.033	0.036	0.046	0.037	95.9	98.3	95.8	98.3
Mean 1–7	0.019	0.017	0.021	0.019	90.1	91.6	89.2	91.1

HSVD-AR with an average $nRMSE$ of 0.019 and an average $mNSE$ of 89.2%. The lowest accuracy was reached by HSVD-ANN with an average $nRMSE$ of 0.047 and an average $mNSE$ of 72.7%. Those blank spaces corresponds to poor results that were obtained.

Table 3.3, Figs. 3.11 and 3.12 present the prediction results of multi-week ahead prediction of Injured-G2 via HSVD-AR, SSA-AR, HSVD-ANN and SSA-ANN. The results shows that the accuracy decreases as the time horizon increases. The best accuracy was reached by using SSA-AR for 1 to 10-week ahead forecasting while for 11 to 14-week ahead forecasting HSVD-AR is more accurate. SSA-AR reaches an average $nRMSE$ of 0.017 and an average $mNSE$ of 94.3%, it is followed by HSVD-AR with an average $nRMSE$ of 0.023 and an average $mNSE$ of 92.1%. The lowest accuracy was reached by HSVD-ANN with an average $nRMSE$ of 0.035 and an average $mNSE$ of 88.1%. Those blank spaces corresponds to poor results that were obtained.

Table 3.4, Figs. 3.13 and 3.14 present the prediction results of multi-week ahead prediction of Injured-G3 via HSVD-AR, SSA-AR, HSVD-ANN and SSA-ANN. The best accuracy was reached by using SSA-AR for 1 to 5-week ahead forecasting while for 6 to 14-week ahead forecasting HSVD-AR is more accurate. SSA-AR reaches an average $nRMSE$ of 0.003 and an average $mNSE$ of 82.7%, it is followed by HSVD-AR with an average $nRMSE$ of 0.01 and an average $mNSE$ of 82.9%. The lowest accuracy was reached by HSVD-ANN with an average $nRMSE$ of

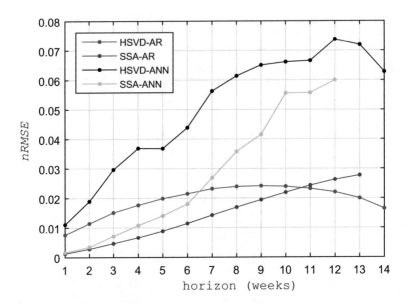

Fig. 3.9 nRMSE results for multi-week ahead forecasting of Injured-G1

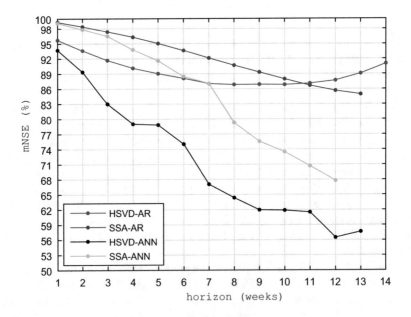

Fig. 3.10 mNSE results for multi-week ahead forecasting of Injured-G1

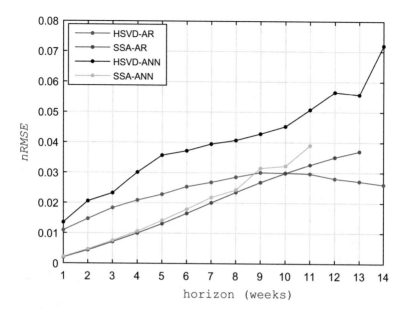

Fig. 3.11   nRMSE results for multi-week ahead forecasting of Injured-G2

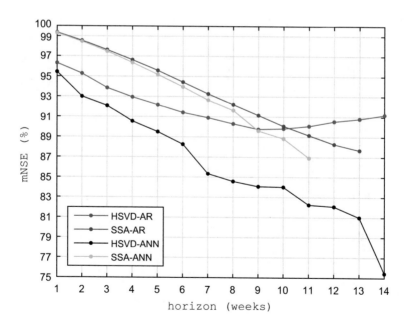

Fig. 3.12   mNSE results for multi-week ahead forecasting of Injured-G2

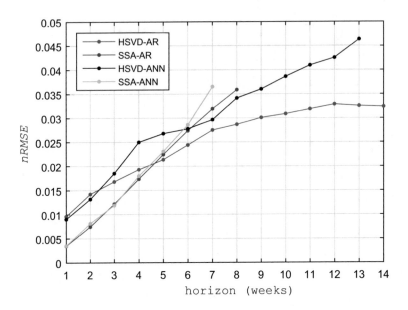

Fig. 3.13 nRMSE results for multi-week ahead forecasting of Injured-G3

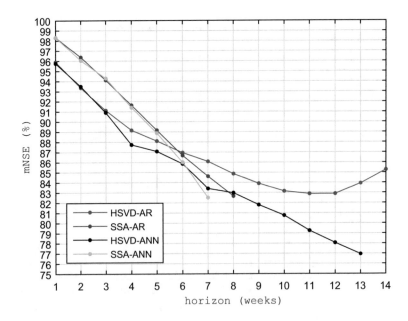

Fig. 3.14 mNSE results for multi-week ahead forecasting of Injured-G3

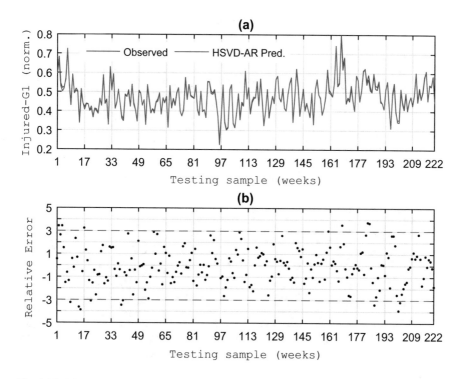

Fig. 3.15 Injured-G1 14-weeks ahead prediction (**a**) Observed versus HSVD-AR prediction (**b**) Relative error

0.009 and an average $mNSE$ of 77.0%. Those blank spaces corresponds to poor results that were obtained.

The results obtained with the computation of the last prediction horizon by means of HSVD-AR are illustrated in Figs. 3.15, 3.16, and 3.17 for Injured-G1, Injured-G2, and Injured-G3 respectively.

Figure 3.15a presents the observed and the predicted signal via HSVD-AR for 14-weeks-ahead prediction of *Injured-G1*. Whereas Fig. 3.15b show the Relative Error. From figures, good fit is observed, HSVD-AR reaches 92.3% of predicted points with a Relative Error lower than ±3%.

The estimation curves for 14-weeks ahead prediction of Injured-G2 and Injured-G3 via HSVD-AR are presented in Figs. 3.16 and 3.17 respectively. High accuracy is observed for both signals. For Injured-G1 91.9% of predicted points present a Relative Error lower than ±5%. While for Injured-G2 94.7% of predicted points present a Relative Error lower than ±6%.

On the other hand, Figs. 3.18, 3.19, and 3.20 present the estimation curves via SSA-AR for 13-week-ahead forecasting. Figure 3.18a show the observed versus SSA-AR prediction, good fit is observed; 91.4% of predicted points show a Relative Error lower than ±5%. The estimation curves for 13-weeks ahead prediction

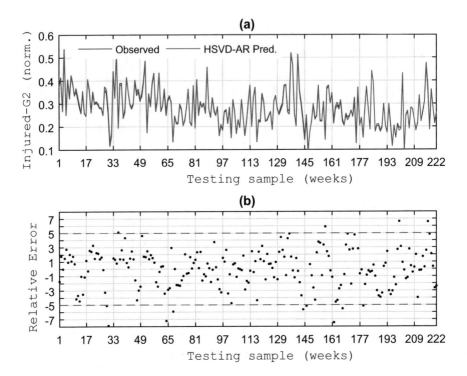

Fig. 3.16 Injured-G2 14-weeks ahead prediction (**a**) Observed versus HSVD-AR prediction (**b**) Relative error

of Injured-G2 and 8-weeks ahead prediction of Injured-G3 via HSVD-AR are presented in Figs. 3.19 and 3.20 respectively. High accuracy is observed for both signals. For Injured-G2 90.1% of predicted points present a Relative Error lower than ±7%. While for Injured-G3 91.6% of predicted points present a Relative Error lower than ±6%.

Conventional forecasting for multiple horizon via conventional AR and conventional ANN was not presented by the poor results that were obtained.

From Tables and Figures presented in this empirical application, it was observed that the decomposition methods HSVD and SSA improve the accuracy of linear and nonlinear models, however the highest improvement was observed in linear models.

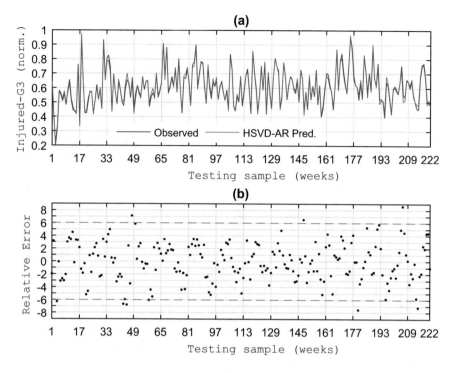

Fig. 3.17 Injured-G3 14-weeks ahead prediction (**a**) Observed versus HSVD-AR prediction (**b**) Relative error

3.7 Empirical Application: Multi-Step Ahead Forecasting of Anchovy and Sardine Fisheries Based on HSVD and SSA via MIMO Strategy

3.7.1 Background

Government institutions, industries and business focus their attention in time series analysis for planning and management activities. Forecasting provides valuable information about the future state of observed variables; unfortunately most of implementations are limited by the high data oscillation. Environmental conditions, economy variables, risk conditions, among others are originated in dynamic systems, consequently their analysis becomes complex and inaccurate results have been often observed.

The fisheries monitoring and management systems can be supported by stock prediction of marine species, however there are limited researching about this topic. Most works present classification and prediction via probabilistic models [24–26]. Prediction based on wavelet and regression was presented in [27] for anchovy, while for sardine [28] presents an ANN-based model.

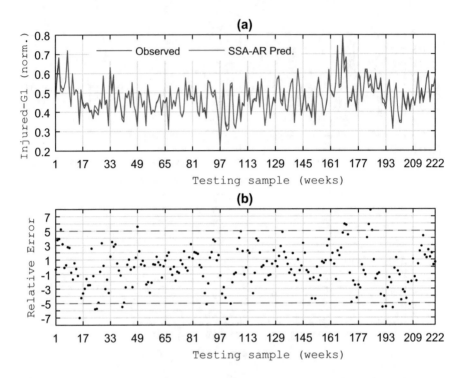

Fig. 3.18 Injured-G1 13-weeks ahead prediction (**a**) Observed versus SSA-AR prediction (**b**) Relative error

In Chile sardine and anchovy are species of great interest, the government institutions make monitoring and researching to assess their sustainable exploitation.

The performance of the models HSVD-AR and SSS-AR is evaluated with monthly series of anchovy and sardine catches during the last five decades. Multi-step ahead forecasting is performed via MIMO strategy.

3.7.2   Models in Evaluation

The same structures that were evaluated in previous Empirical Application are used here. Figure 3.1, shows the stages of preprocessing and prediction. In the preprocessing stage are extracted the intrinsic components of low and high frequency. Whereas in the prediction the components are used as inputs of the model, in this case only the AR model will be implemented. After, the MIMO strategy is developed to extend the forecast horizon.

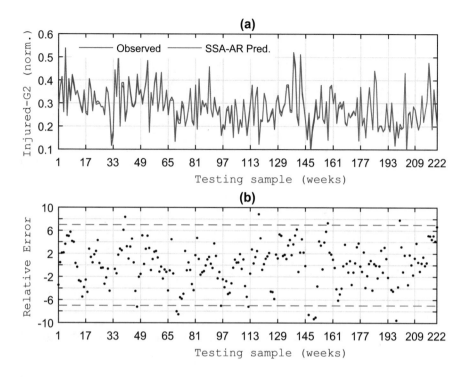

Fig. 3.19 Injured-G2 13-weeks ahead prediction (**a**) Observed versus SSA-AR prediction (**b**) Relative error

3.7.3   Data Description

The stock of anchovy and sardine is in dependence of the environmental conditions; significant fluctuations of the marine ecosystem causes high stock variability. In this case study, the monthly nonstationary time series of anchovy and sardine catches are used, the data were collected by the National Service of fisheries and aquaculture (SERNAPESCA) [29], along the Chilean southern coast (33°S–36°S), from January 1958 for anchovy and from 1949 for sardine, to December 2011.

Considering the information of SERNAPESCA, it is observed that during the last decade the anchovy stock had been decreasing, while the sardine stock had been increasing; it is advisable to keep track of these species and their evolution, which is essential for exploitation policies and control rules.

Figures 3.21a and 3.22a show the observed time series, whereas Figs. 3.21b and 3.22b show the Fourier Power Spectrum (FPS) of anchovy and sardine respectively. High data variability is observed during the analyzed period; For example anchovy presents uptrend until year 2008 followed by downtrend until end; whereas sardine shows scarcity from 1949 to 1989, followed by uptrend and abundance in the last years.

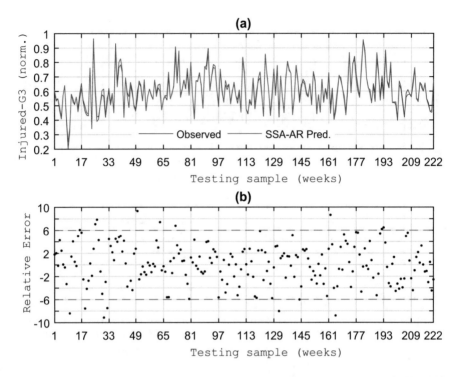

Fig. 3.20 Injured-G3 8-weeks ahead prediction (**a**) Observed versus SSA-AR prediction (**b**) Relative error

The FPS analysis shows the signal spectrum and the red noise spectrum, a signal spectrum peak is significative when its value is higher than the red noise spectrum [22]. Both series present the shortest significant peak of low frequency at month 30 at 65% of confidence level; besides 30 is a multiple harmonic of high frequency. The peaks information is used to set the AR model order ($P = 30$).

3.7.4   Forecasting

3.7.4.1   Decomposition in Intrinsic Components

The intrinsic components are extracted from the time series through HSVD and SSA. The differential energy of the singular values informed about the effective window length, in this case $L = 16$ for both, anchovy and sardine. The embedding matrix H of $L \times K$ dimension is decomposed in singular values and singular vectors. Elemental matrix H_1 was computed to extract the low frequency component C_L via unembedding or diagonal averaging for HSVD or SSA respectively.

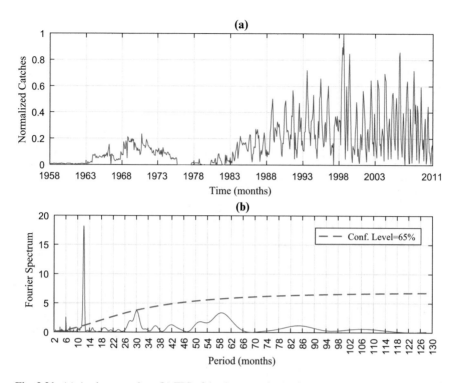

Fig. 3.21 (**a**) Anchovy catches, (**b**) FPS of Anchovy catches series

The components of low and high frequency extracted from *Anchovy* are shown in Fig. 3.23. Whereas the extracted from *Sardine* are shown in Fig. 3.24.

Long-memory periodicity features were observed in all C_L components, whereas short-term periodic fluctuations were identified in all C_H components.

3.7.4.2 Prediction Based on Intrinsic Components

Each data set of low and high frequency that was obtained in the preprocessing stage has been divided into two subsets, training and testing for prediction. The training subset involves 80% of the samples, and consequently the testing subset involves the remaining 20%. The prediction is implemented by means of the autoregressive model, linear (AR) via MIMO strategy.

Table 3.5 shows the prediction results for multi-step ahead prediction of Anchovy and Sardine.

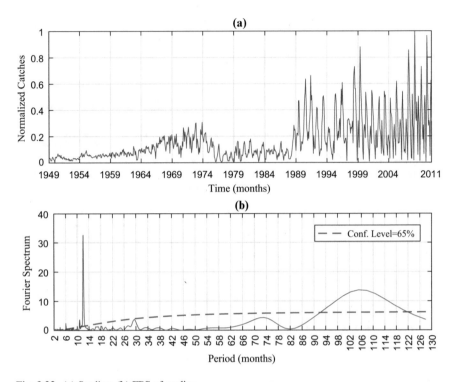

Fig. 3.22 (a) Sardine, (b) FPS of sardine

Table 3.5, Figs. 3.25 and 3.26, present the prediction results of multi-month ahead prediction of Anchovy via HSVD-AR and SSA-AR. The results show that the accuracy decreases as the time horizon increases. The best accuracy was reached by using SSA-AR for 1 to 12-months ahead forecasting while for 13 to 15-months ahead forecasting, HSVD-AR is more accurate. SSA-AR present an average $nRMSE$ of 0.024 and an average $mNSE$ of 89.1% whereas HSVD-AR presents an average $nRMSE$ of 0.016 and an average $mNSE$ of 92.2%. Those blank spaces corresponds to poor results that were obtained.

Table 3.6, Figs. 3.27 and 3.28, present the prediction results of multi-month ahead prediction of Sardine via HSVD-AR and SSA-AR. The best accuracy was reached by using SSA-AR for 1 to 12-months ahead forecasting while for 13 to 15-months ahead forecasting, HSVD-AR is more accurate. SSA-AR present an average $nRMSE$ of 0.014 and an average $mNSE$ of 92.9% whereas HSVD-AR presents an average $nRMSE$ of 0.018 and an average $mNSE$ of 92.3%. Those blank spaces corresponds to poor results that were obtained.

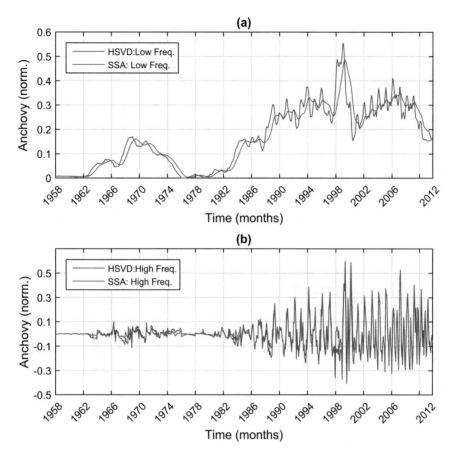

Fig. 3.23 Anchovy components (**a**) Low frequency components via HSVD and SSA, (**b**) High frequency components via HSVD and SSA

The results of the last prediction horizon by means of HSVD-AR are illustrated in Figs. 3.29 and 3.30 for anchovy and sardine respectively. Figure 3.29a presents the observed and the predicted signal via HSVD-AR for 14-weeks-ahead prediction of *Anchovy*, whereas Fig. 3.29b show the Linear fit. Figure 3.30a presents the observed and the predicted signal via HSVD-AR for 13-weeks-ahead prediction of *Sardine*, whereas Fig. 3.30b show the Linear fit.

The results obtained with the computation of the last prediction horizon by means of SSA-AR are illustrated in Figs. 3.31 and 3.32 for anchovy and sardine respectively. Figure 3.31a presents the observed and the predicted signal via SSA-AR for 14-weeks-ahead prediction of *Anchovy*, whereas Fig. 3.31b show the Linear fit. Figure 3.32a presents the observed and the predicted signal via SSA-AR for 13-weeks-ahead prediction of *Sardine*, whereas Fig. 3.32b show the Linear fit.

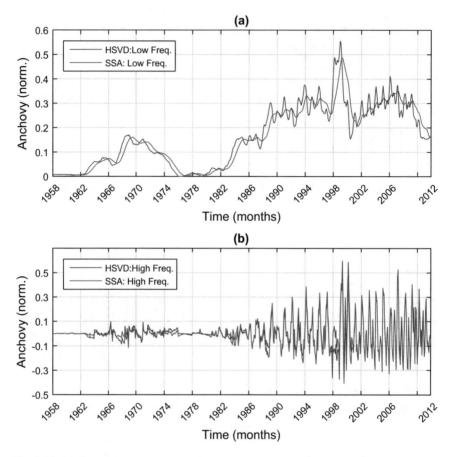

Fig. 3.24 Sardine components (**a**) Low frequency components via HSVD and SSA, (**b**) High frequency components via HSVD and SSA

3.8   Chapter Conclusions

Singular Value Decomposition of a Hankel matrix and Singular Spectrum Analysis were implemented under similar criteria. HSVD and SSA depends of the selection of an effective window length L to extract the components of low and high frequency from a nonstationary time series. The computation of the energy of the singular values gives some information to orient the selection of L. Some experiments were developed examining diverse effective window length values, finally it was chosen by trial and error from those candidate values that were given by the singular values differential energy.

Table 3.5 Multi-step ahead forecasting results for *Anchovy* via MIMO strategy

h (month)	nRMSE		mNSE (%)	
	HSVD-AR	SSA-AR	HSVD-AR	SSA-AR
1	0.006	0.001	97.2	99.6
2	0.011	0.002	95.0	98.9
3	0.014	0.004	93.4	98.0
4	0.018	0.007	91.5	96.8
5	0.022	0.009	89.6	95.5
6	0.025	0.012	88.2	94.1
7	0.027	0.015	87.6	92.8
8	0.028	0.018	87.8	91.4
9	0.029	0.020	87.7	90.2
10	0.028	0.023	87.6	89.0
11	0.029	0.025	87.0	87.5
12	0.030	0.028	85.9	86.4
13	0.031	0.030	86.0	85.6
14	0.032	0.032	86.1	84.6
15	0.031		86.3	
Min	0.006	0.001	85.9	84.6
Max	0.032	0.032	97.2	99.6
Mean	0.024	0.016	89.1	92.2

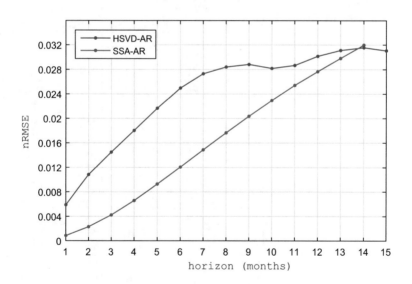

Fig. 3.25 nRMSE results for multi-week ahead forecasting of Anchovy

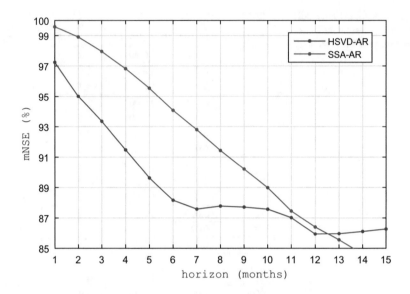

Fig. 3.26 mNSE results for multi-week ahead forecasting of Anchovy

Table 3.6 Multi-step ahead forecasting results for *Sardine* via MIMO strategy

h (month)	nRMSE		mNSE (%)	
	HSVD-AR	SSA-AR	HSVD-AR	SSA-AR
1	0.005	0.001	97.5	99.6
2	0.009	0.002	95.8	98.8
3	0.012	0.004	94.5	98.0
4	0.014	0.006	93.2	97.0
5	0.017	0.008	91.8	95.8
6	0.019	0.011	90.9	94.5
7	0.019	0.014	90.7	93.3
8	0.020	0.016	90.7	92.1
9	0.022	0.018	89.7	90.9
10	0.022	0.021	89.5	89.8
11	0.021	0.023	89.9	88.7
12	0.021	0.024	89.6	88.2
13	0.024	0.026	88.7	87.3
14	0.024	0.028	88.2	86.3
15	0.023		88.6	
Min	0.005	0.001	88.2	86.3
Max	0.024	0.028	97.5	99.5
Mean	0.018	0.014	91.3	92.9

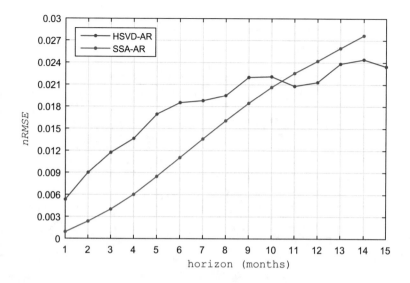

Fig. 3.27   nRMSE results for multi-week ahead forecasting of Sardine

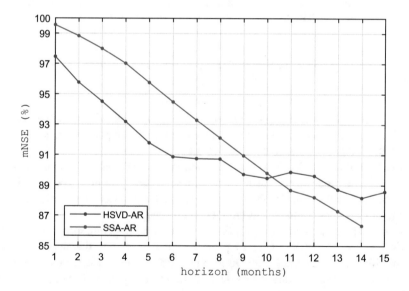

Fig. 3.28   mNSE Results for multi-week ahead forecasting of Sardine

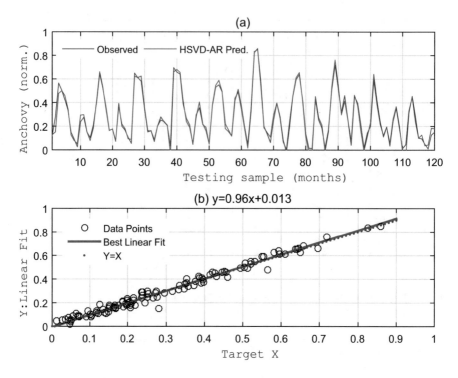

Fig. 3.29 Anchovy 15-month ahead prediction (**a**) Observed versus HSVD-AR prediction (**b**) Linear fit error

In Chap. 3, it was found that an effective window length of $L = 2$ is enough to obtain appropriate decomposition useful for one-step ahead prediction, however for multi-step ahead forecasting the accuracy decreases.

Regarding the prediction, it was observed that HSVD and SSA improve linear and nonlinear models, but the highest gain in accuracy is observed with linear models. This conclusion is validated by means of the empirical applications, which was found in previous chapter and ratified in this chapter.

Multi-step ahead forecasting based on HSVD and SSA combined with AR and ANN via direct or MIMO strategy present high accuracy. In this approach there is no difference in accuracy with Direct and MIMO strategies when the prediction model is AR. The advantage of MIMO over Direct strategy is in the simplicity of an unique model.

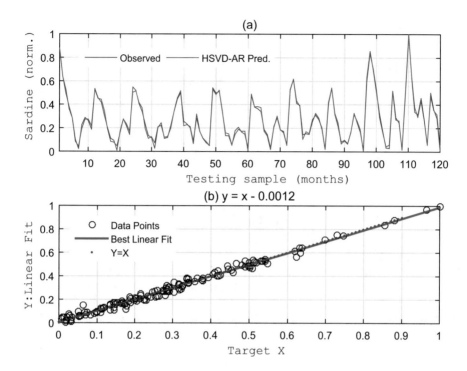

Fig. 3.30 Sardine 15-month ahead prediction (**a**) Observed versus HSVD-AR prediction (**b**) Linear fit error

The best mean accuracy was observed with SSA-AR for nearest horizons, whereas for farthest horizons HSVD-AR is more accurate. By instance, multi-step ahead forecasting of Injured-G1 by means of HSVD-AR reaches a mean $mNSE$ of 89.2% (for 14-weeks), and a $mNSE$ of 92.7% through SSA-AR (for 13-weeks), which means that SSA-AR present a 4% of gain in MAPE with respect to HSVD-AR. The average gain (considering all series) in $mNSE$ is of SSA-AR with respect to HSVD-AR is of 2.7%. The average gain in $mNSE$ of linear models with respect to nonlinear models for multi-step ahead forecasting of traffic accidents is of 5.8%. And the gain of models based on decomposition in intrinsic components of low and high frequency with respect to conventional models is extreme and incalculable due to those models present comparable results only for one-step ahead prediction.

Regarding the analyzed problems. For traffic accidents, were obtained 14-weeks ahead forecasting of series that characterize the most relevant causes of incidents with presence of injured. Whereas for fisheries of anchovy and sardine were obtained 15-months ahead forecasting from 30 observations. These results are relevant for annual management and planning.

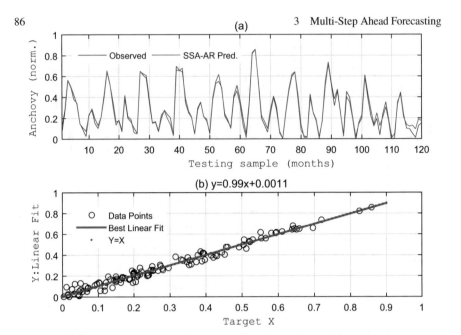

Fig. 3.31 Anchovy 14-month ahead prediction (**a**) Observed versus SSA-AR prediction (**b**) Linear fit error

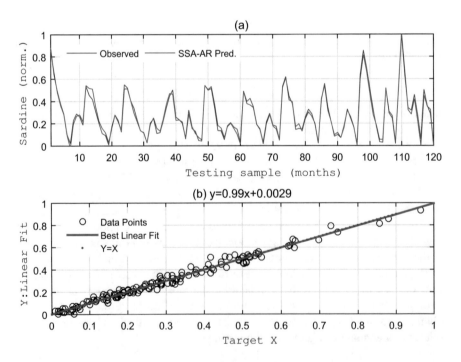

Fig. 3.32 Sardine 14-month ahead prediction (**a**) Observed versus SSA-AR prediction (**b**) Linear fit error

References

1. Weigend AS, Gershenfeld NA (1994) Time series prediction: prediction the future and understanding the past. Santa Fe Institute Studies in the Sciences of Complexity, Proceedings Vol XV, Reading, MA
2. Tiao GC, Tsay RS (1994) Some advances in non-linear and adaptive modelling in time-series. J Forecast 13:109–131
3. Chen R, Yang L, Hafner C (2004) Nonparametric multistep-ahead prediction in time series analysis. J R Stat Soc B 66:669–686
4. Andalib A, Atry F (2009) Multi-step ahead forecasts for electricity prices using NARX: a new approach, a critical analysis of one-step ahead forecasts. Energy Convers Manag 50(3):739–747
5. Xiong T, Bao Y, Hu Z (2013) Beyond one-step-ahead forecasting: evaluation of alternative multi-step-ahead forecasting models for crude oil prices. Energy Econ 40:405–415
6. Sorjamaa A, Hao J, Reyhani N, Ji Y, Lendasse A (2007) Methodology for long-term prediction of time series. Neurocomputing 70(16):2861–2869
7. Shi Z, Han M (2007) Support vector echo-state machine for chaotic time-series prediction. IEEE Trans Neural Netw 18(2):359–372
8. Hamzaçebi C, Akay D, Kutay F (2009) Comparison of direct and iterative artificial neural network forecast approaches in multi-periodic time series forecasting. Expert Syst Appl 36:3839–3844
9. Atiya AF, and El-Shoura SM, Shaheen SI, El-Sherif MS (1999) A comparison between neural network forecasting techniques, case study: river flow forecasting. IEEE Trans Neural Netw 10(2):402–409
10. De Giorgi MG, Malvoni M, Congedo PM (2016) Comparison of strategies for multi-step ahead photovoltaic power forecasting models based on hybrid group method of data handling networks and least square support vector machine. Energy 107:360–373
11. Wang J, Song Y, Liu F, Hou R (2016) Analysis and application of forecasting models in wind power integration: a review of multi-step-ahead wind speed forecasting models. Renew Sust Energy Rev 60:960–981
12. Broomhead DS, King GP (1986) Extracting qualitative dynamics from experimental data. Phys D Nonlinear Phenom 20(2):217–236
13. Golyandina N, Nekrutkin V, Zhigljavsky A (2001) Analysis of time series structure: SSA and related techniques. Chapman & Hall/CRC, Boca Raton, pp 15–44
14. Viljoen H, Nel DG (2010) Common singular spectrum analysis of several time series. J Stat Plan Inference 140(1):260–267
15. Chen Q, van Dam T, Sneeuw N, Collilieux X, Weigelt M, Rebischung P (2013) Singular spectrum analysis for modeling seasonal signals from GPS time series. J Geodyn 728:25–35
16. Muruganatham B, Sanjith MA, Krishnakumar B, Satya Murty SAV (2013) Roller element bearing fault diagnosis using singular spectrum analysis. Mech Syst Signal Process 35(1):150–166
17. Elsner JB, Tsonis AA (1996) Singular spectrum analysis: a new tool in time series analysis. Plenum Press, New York
18. Hassani H, Mahmoudvand R, Zokaei M (2011) Separability and window length in singular spectrum analysis. C R Math 349(17):987–990
19. Wang R, Ma HG, Liu GQ, Zuo DG (2015) Selection of window length for singular spectrum analysis. J Frankl Inst 352:1541–1560
20. Freeman JA, Skapura DM (1991) Neural networks, algorithms, applications, and programming techniques. Addison-Wesley, Boston
21. CONASET (2015) Comisión Nacional de Seguridad de Tránsito. http://www.conaset.cl
22. Torrence C, Compo GP (1998) A practical guide to wavelet analysis. Bull Am Meteorol Soc 79(1):61–78

23. Kwiatkowski D, Phillips PCB, Schmidt P, Shin Y (1992) Testing the null hypothesis of stationarity against the alternative of a unit root. J Econ 54(1):159–178
24. Varis O, Kuikka S (1997) Joint use of multiple environmental assessment models by a Bayesian meta-model: the Baltic salmon case. Ecol Model 102:341–351
25. Lee DC, Rieman BE (1997) Population viability assessment of Salmonids by using probabilistic networks. N Am J Fish Manag 17:1144–1157
26. Fernandes JA, Irigoien X, Lozano JA, Inza I, Goikoetxea N, Pérez A (2015) Evaluating machine-learning techniques for recruitment forecasting of seven North East Atlantic fish species. Eco Inform 25:35–42
27. Rodríguez N, Cubillos C, Rubio JM (2014) Multi-step-ahead forecasting model for monthly anchovy catches based on wavelet analysis. J Appl Math 2014:8 p, Article ID 798464
28. Yáñez E, Plaza F, Gutiérrez-Estrada JC, Rodríguez N, Barbieri MA, Pulido-Calvo I, Bórquez C (2010) Anchovy (Engraulis ringens) and sardine (Sardinops sagax) abundance forecast off northern Chile: a multivariate ecosystemic neural network approach. Prog Oceanogr 87(1):242–250
29. SERNAPESCA, Servicio Nacional de Pesca y Acuicultura. http://www.sernapesca.cl/.Visited 12/2013

Chapter 4
Multilevel Singular Value Decomposition

4.1 Forecasting Methodology

The proposed forecasting methodology is described in two stages, preprocessing and prediction as it is shown in Fig. 4.1. In the preprocessing stage the time series is decomposed in two additive components of low and high frequency via MSVD. Whereas in the prediction stage is implemented the Autoregressive model via Direct or MIMO strategy.

In the preprocessing stage is also used Stationary Wavelet Transform with comparison purpose.

4.1.1 Multilevel Singular Value Decomposition

MSVD method is inspired in the pyramidal process implemented in multiresolution analysis of Mallat Algorithm [1] which was defined for wavelet representation. This method, at difference of HSVD, implements iterative embedding and pyramidal decomposition with a fixed window length $L = 2$ [2].

MSVD algorithm is summarized as the pseudo code shown in Fig. 4.2. The input algorithm is the observed time series x of length N, and at the end, two additive and intrinsic components are obtained as outputs, c_L and c_H, which represent the low frequency and the high frequency component respectively, each one of length N. MSVD is performed in three steps, *embedding* through a Hankel matrix of $2 \times (N - 1)$ dimension as (4.2), *decomposition* in orthogonal matrices of eigenvectors U and V, and singular values λ_1, λ_2; finally *unembedding* from elementary matrices H_1 and H_2.

MSVD is processed iteratively and it is controlled by J until the k repetition is done. When the singular spectrum rate ΔR reaches the asymptotic point the k value is set. The computation of ΔR is made by means of the next equations:

© Springer International Publishing AG, part of Springer Nature 2018
L. M. Barba Maggi, *Multiscale Forecasting Models*,
https://doi.org/10.1007/978-3-319-94992-5_4

Fig. 4.1 Forecasting models based on MSVD and SWT

Normalize time series $X = X./max(abs(X))$
Set the counter $J = 0$, and signal to be decomposed $A_0 = X$
while $(J < k)$
$J = J + 1$
Embed the signal H= $hankel(A_0, 2)$
Decompose the matrix U,[λ_1 λ_2],V]=svd(H)
Compute elementary matrix 1: $\mathrm{H}_1 = U_1 \lambda_1 V_1'$
Compute elementary matrix 2: $\mathrm{H}_2 = U_2 \lambda_2 V_2'$
Extract the low frequency signal of level J,
$A_J = [\mathrm{H}_1(1, 1 : end) \quad \mathrm{H}_1(2, end)]$
Extract the high frequency signal of level J,
$d_J = [\mathrm{H}_2(1, 1 : end) \quad \mathrm{H}_2(2, end)]$
Update the decomposition signal $A_0 = A_J$
end while
Get the low frequency component $C_L = A_J$
Get the high frequency component $C_H = \Sigma_{i=1}^{J} d_i$

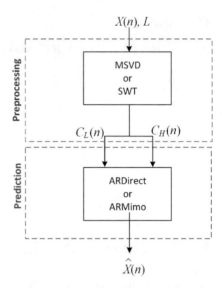

Fig. 4.2 Multilevel SVD algorithm

$$\Delta R_j = \frac{R_j}{R_{j-1}} \tag{4.1a}$$

$$R_j = \frac{\lambda_{1,j}}{\lambda_{1,j} + \lambda_{2,j}} \tag{4.1b}$$

where R_j is the Relative Energy of the singular values, and $j = 1, 2$.

The Hankel matrix has the form:

$$H = \begin{pmatrix} x_1 & x_2 & \dots & x_{N-1} \\ x_2 & x_{2+1} & \dots & x_N \end{pmatrix}.$$ (4.2)

The Hankel matrix H in all repetitions will have $2 \times K$ dimension.

4.1.2 Multi-Step Ahead Forecasting

The implementation of multi-step ahead forecasting is based on the AR model and MIMO strategy. The inputs of the forecasting model are the components of low and high frequency extracted through MSVD.

MIMO computes the output for the forecast horizon in a single simulation with an unique model. The output is a vector rather than a scalar value. Consider:

$$\left[\hat{X}(n+1), \hat{X}(n+2), \dots, \hat{X}(n+\tau) \right] = f\left[Z(n), Z(n-1), \dots, Z(n-P+1) \right],$$ (4.3)

where n is the time instant, τ is the maximum forecast horizon, Z_i is ith input regressor vector of size $2P$ which contains the lagged values of C_L and C_H, each one of size P.

The following equation defines MIMO in matrix form:

$$\hat{X} = \beta Z^T,$$ (4.4)

where β is a $\tau \times 2P$ matrix of linear coefficients and Z^T is a $N_{te} \times 2P$ transposed regressor matrix, used for testing. The regressor matrix is built with P lagged values of each component, it has the form

$$Z = (C_L((n), C_L(n-1), \dots, C_L(n-P+1), C_H(n), C_H(n-1), \dots, C_H(n-P+1).$$ (4.5)

The matrix of coefficients β is computed with the Least Square Method (LSM) as below:

$$\beta = XZ^{\dagger}$$ (4.6)

where $Z^{\dagger}$ is the Moore-Penrose pseudoinverse matrix, and X is a $N_{tr} \times 2P$ matrix of observed values used for training.

4.2 Wavelet Theory

Wavelet theory was initiated by the work of Grossman and Morlet in 1984 for geophysical signal processing [3]. Multiscale view of signal analysis is in the essence wavelet transform. In wavelet a signal is analyzed at different scales or resolutions which is of great help to locate finely long-term behavior in nonstationary time series.

The basis for wavelet transform is wavelet function. The wavelet is a quickly vanishing oscillating function localized both in frequency and time. In both forms of wavelet analysis (continuous and discrete), a signal is decomposed into scaled and translated versions of a single function ψ_t called *mother wavelet*:

$$\psi_t(a, b) = \frac{1}{\sqrt{|a|}} \psi\left(\frac{t - b}{a}\right), \tag{4.7}$$

where a and b are the scale and translation parameters, respectively, with $a, b \in \Re$ and $a \neq 0$.

The *continuous wavelet transform* (CT) of a signal $X(t) \in L^2(\Re)$ (the space of the square integrable functions) is defined as

$$C(a, b) = \int_{-\infty}^{+\infty} X(t) \frac{1}{\sqrt{|a|}} \psi * \left(\frac{t - b}{a}\right) dt, \quad = \langle X(t), \psi_t(a, b) \rangle, \tag{4.8a}$$

where the symbol * means complex conjugations and $\langle \rangle$ the inner product.

The Discrete Wavelet Transform (DWT) is obtained by discretizing the parameters a and b. In most common form DWT employs a dyadic sampling with parameters a and b based on powers of two: $a = 2^j$, $b = k2^j$, with $k \in Z$. The dyadic wavelets are

$$\psi_t(j, k) = 2^{(-j/2)} \psi(2^{-j}t - k). \tag{4.9}$$

The DWT can be written as

$$d(j, k) = \int_{-\infty}^{+\infty} X(t) 2^{(-j/2)} \psi * (2^{-j}t - k) dt, \quad = \langle X(t), \psi_t(j, k) \rangle, \tag{4.10a}$$

where $d(j, k)$ are known as wavelet (or detail) coefficients at scale j and location k.

Mallat in 1989 developed Multiresolution analysis for implementing Discrete Wavelet Transform (DWT) through a filter bank to separate signals [1]. The multiresolution decomposition of a signal $X(t)$ is expressed with

$$X(t) = \sum_{j=-\infty}^{+\infty} \sum_{k=-\infty}^{+\infty} d(j, k) 2^{-j/2} \psi(2^{-j}t - k). \tag{4.11}$$

The signal decomposition can be efficiently implemented through iterative low-pass and high-pass filtering combined with down-sampling operations. The high-pass filters are associated with the wavelet functions $\psi_t(j, k)$, while the low-pass filters are associated with the scaling functions $\varphi_t(j, k)$. At each step, the pair of filters decomposes the signal into low-frequency components (approximation coefficients), and high-frequency components (details coefficients). Filtering is applied first to the original signal, and then recursively, to the approximation series only. At every iteration, the output of each filter is down-sampled by a factor of 2 (decimation) halving the data each time. The down-sampling accounts for the speed of the algorithm, reducing the computation at each iteration geometrically (after J iteration the number of samples being manipulated shrinks by 2^J) [4].

The simplest example of wavelet filter is Haar [5]. Orthonormal wavelets known as Daubechies are very popular, they are based on iterations of discrete filters [6].

Wavelet functions frequently used are Haar, Daubechies, Coiflets, Symlets and Meyer. Distinctive features characterize a wavelet function including their region of support and the corresponding number of vanishing moments [7]. The wavelet support region refers to the span length of the given wavelet function which affects its feature localization abilities. Whereas the vanishing moment limits refers to the ability for representing polynomial behavior or data information. For example, daubechies, coiflets, and symlets with one moment (db1,coif1,sym1), encrypts one coefficient polynomials, or constant signal components. Wavelet functions with two moments encrypts two coefficient, etc. Daubechies [8] and Addison [9] present more details on wavelet families.

DWT is computed only on a dyadic grid of points, where the downsampling is at different rate for different scales. DWT needs that a sample of size N to be an integer multiple of 2^J, and the number of N_j, of scaling and wavelet coefficient at each level decreases by a factor of two, due to the decimation process that need to be applied at the output of the corresponding filters [10].

Classical DWT is therefore characterized by downsampling which induces *translation-variance*. This makes critical when the decomposition is used in forecasting.

The coefficients of the DWT can be of approximation and detail. Approximation coefficient represents the high-scale and low-frequency components of the signal, and the other is the detail coefficient represents the low-scale and high-frequency components of the signal.

Since DWT is an orthogonal transform, it corresponds to a particular choice of basis for the space $\mathbb{R}^N$ in which the original sequence lies. Several possible modifications of DWT can be selected, each corresponding to a somewhat different choice of basis.

Undecimated DWT also known as Oversampled Wavelet Transform is a type of Discrete Wavelet Transform with translation-invariance. Stationary Wavelet Transform (SWT) gets translation-invariant avoiding the downsampling operation. In SWT the decomposed signal is never sub-sampled and instead, the signal is up-sampled at each level of decomposition. The signal is undecimated, and the two new sequences at each level have the same length as the original signal. In the practice, the filters are modified at each level by padding them out with zeros [11].

4.2.1   Stationary Wavelet Transform

SWT decomposition is illustrated in Fig. 4.3. SWT is based on the Discrete Wavelet Transform which implementation is defined in the algorithm of Shensa [12]. SWT implements up-sampled filtering [11, 13].

In SWT the length of the observed signal must be an integer multiple of 2^j, with $j = 1, 2, \ldots, J$; where J is the scale number. The signal is separated in approximation coefficients and detail coefficients at different scales, this hierarchical process is called multiresolution decomposition [1].

The observed signal a_0 (which was named x in previous section) is decomposed in approximation and detail coefficients through a bank of low pass filters $(h_0, h_1, \ldots, h_{J-1})$ and a bank of high pass filter $(g_0, g_1, \ldots, g_{J-1})$ one to each level as the scheme of Fig. 4.3. Each level filters are up-sampled versions of the previous ones. The components obtained after the decomposition are never decimated, therefore they have the same length as the observed signal.

At first decomposition level the observed signal a_0 is convoluted with the first low pass filter h_0 to obtain the first approximation coefficients a_1 and with the first high pass filter g_0 to obtain the first detail coefficients d_1. The process is defined as follows

$$a_1(n) = \sum_i h_0(i)a_0(n - i), \tag{4.12a}$$

$$d_1(n) = \sum_i g_0(i)a_0(n - i), \tag{4.12b}$$

The process follows iteratively, for $j = 1, \ldots, J - 1$ and it is given as

$$a_{j+1}(n) = \sum_i h_j(i)a_j(n - i), \tag{4.13a}$$

$$d_{j+1}(n) = \sum_i g_j(i)a_j(n - i), \tag{4.13b}$$

Inverse Stationary Wavelet Transform (iSWT) performs the reconstruction. The implementation of iSWT consists in applying the operations that were performed

Fig. 4.3 Decomposition scheme of SWT with $J = 3$

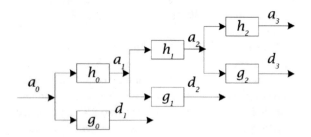

in SWT but in inverse order and based on equivalent filters of reconstruction. SWT obtains sub bands of frequency, the last approximation coefficient reconstructed $\tilde{a}_J$ gives rise the component of low frequency c_L, whereas all reconstructed detail coefficients $\tilde{d}_j$ are added to obtain the component of high frequency c_H.

4.3 Empirical Application: MSVD and SWT for Multi-Step Ahead Traffic Accidents Forecasting

4.3.1 Background

In this empirical application MSVD-MIMO and SWT-MIMO are implemented for multi-step ahead forecasting of traffic accidents.

The hybrid models are evaluated with time series of traffic accidents coming from Santiago which were used in previous chapter. The data were weekly collected during the period 2000:2014 and they characterize improper behavior of drivers, passengers, and pedestrians on roads, and other causes of different nature. Police and government institutions make monitoring and promote prevention activities to encourage responsible attitude of drivers, passengers, and pedestrians by effective road safety policies.

4.3.2 Data Description

Three relevant nonstationary time series of traffic accidents are used, the data were collected by the Chilean police and the National Traffic Safety Commission (CONASET) from year 2000 to 2014 in Santiago de Chile with 783 weekly registers. This problem has socioeconomic relevance in planning and management; besides forecasting based on the principal improper behavior of drivers, passengers, and pedestrians will contribute highly in prevention tasks.

CONASET has defined 100 causes of traffic accidents, in this study case are used the analysis groups that were created previously by Barba and Rodríguez [14]. Injured-G1 involves 10 main causes of traffic accidents related with improper behavior of drivers, passengers, and pedestrians. Injured-G2 involves 10 secondary causes also related with improper behavior. Injured-G3 groups the remaining causes (not related with improper behavior). Detailed information was shown in Table 3.1.

Figures 4.4a, 4.5a, and 4.6a show the observed time series Injured-G1, Injured-G2, and Injured-G3 respectively. Whereas Figs. 4.4b, 4.5b, and 4.6b show their Fourier Power Spectrum (FPS). The FPS analysis shown the signal spectrum and the red noise spectrum, both series, Injured-G1 and Injured-G2 present the highest peak at week 26 at 98% of confidence level, whereas for Injured-G3 shows the highest peak at week 17 at 73% of confidence level. Whit the peaks information the AR order is set, $P = 26$ for Injured-G1 and Injured-G2, and $P = 17$ for Injured-G3.

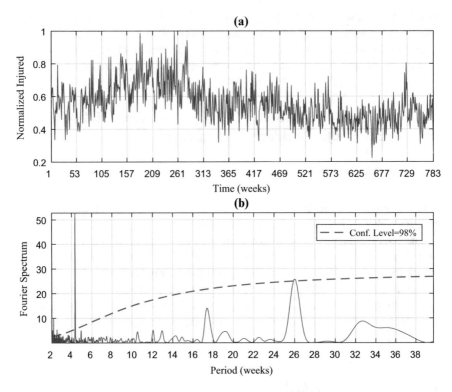

Fig. 4.4 (a) Injured-G1, (b) FPS of Injured-G1. The thick solid line is the global wavelet spectrum for Injured-G1, while the dashed line is the red-noise spectrum

4.3.2.1 Decomposition Based on MSVD and SWT

MSVD implements an iterative process which finish when ΔR reaches the asymptotic value. The Singular Spectrum Rate ΔR for each decomposition level J are illustrated in Fig. 4.7, the asymptotic value is reached when $\Delta R \approx 1$, and it was observed in the repetition 16. Therefore the iterative process finish when iteration 16 was performed, this condition is used with all time series.

SWT decomposition is implemented through Daubechies of order 2 (Db2) (due to the inaccurate results that were obtained with the other type of wavelet functions they are not presented). Three decomposition levels ($J = 3$) were selected according with the period fluctuation between 8 and 16 weeks.

Figures 4.8, 4.9, and 4.10, show the components of low frequency and high frequency obtained with MSVD and SWT for Injured-G1, Injured-G2, and Injured-G3 respectively. The c_L components extracted by both MSVD and SWT, show long-memory periodicity features, whereas the c_H components show short-term periodic fluctuations.

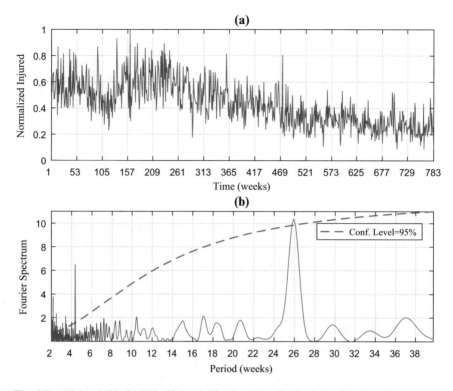

Fig. 4.5 (**a**) Injured-G2, (**b**) FPS of Injured-G2. The thick solid line is the global wavelet spectrum for Injured-G2, while the dashed line is the red-noise spectrum

Before prediction, each data set of low and high frequency has been divided into two subsets, training and testing; the training subset involves 70% of the samples, and consequently the testing subset involves the remaining 30%.

4.3.2.2 Prediction Through MSVD-MIMO and SWT-MIMO

The MIMO model is implemented to predict the number of injured people in traffic accidents for multiple horizon. The Autoregressive model is used with MIMO, the spectral analysis developed through FPS informs about the order of the model, it was shown in Figs. 4.4b, 4.5b, and 4.6b. The inputs of the AR model are the P lagged values of c_L and the P lagged values of c_H, and the outputs are the number of injured for the next τ weeks. The prediction performance is evaluated with efficiency metrics $nRMSE$, $mNSE$, and mIA, which are presented in Tables 4.1, 4.2, and 4.3 for Injured-G1, Injured-G2, and Injured-G3 respectively.

From Table 4.1 and Fig. 4.11, both models present good accuracy, however MSVD-MIMO model presents higher accuracy in comparison with SWT-MIMO.

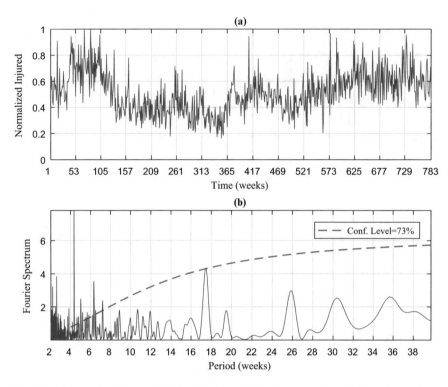

Fig. 4.6 (**a**) Injured-G3, (**b**) FPS of Injured-G3. The thick solid line is the global wavelet spectrum for Injured-G3, while the dashed line is the red-noise spectrum

MSVD-MIMO obtains a significative gain in each forecasting horizon. The mean gain for 1–13 weeks of MSVD-MIMO over SWT-MIMO is 17.7% in $mNSE$ and 8.1% in mIA. SWT-MIMO results for 14-weeks ahead prediction are not presented due to the poor results that were obtained.

The highest accuracy obtained by MSVD-MIMO for Injured-G1 is also obtained for Injured-G2. From Table 4.2 and Fig. 4.12. MSVD-MIMO obtains a significative gain in each forecasting horizon; the mean gain for 1–13 weeks of MSVD-MIMO over SWT-MIMO is 20.6% in $mNSE$ and 9.3% in mIA. SWT-MIMO results for 14-weeks ahead prediction are not presented due to the poor results.

As previous analysis that was made for Injured-G1 and Injured-G2, in the Injured-G3 prediction is also observed the superiority of MSVD-MIMO over SWT-MIMO. From Table 4.3 and Fig. 4.13, MSVD-MIMO obtains a significative gain in each forecasting horizon. The mean gain for 1–13 weeks of MSVD-MIMO over SWT-MIMO is 20.9% in $mNSE$ and 9.4% in mIA. SWT-MIMO results for 14-weeks ahead prediction are not presented due to the poor results.

The Injured-G1 prediction via MSVD-MIMO for 14-weeks ahead prediction is shown in Fig. 4.14a, b; from figures good fit is observed between actual and estimated values. Metrics computation give a $nRMSE$ of 2.9%, a $mNSE$ of 83.3%

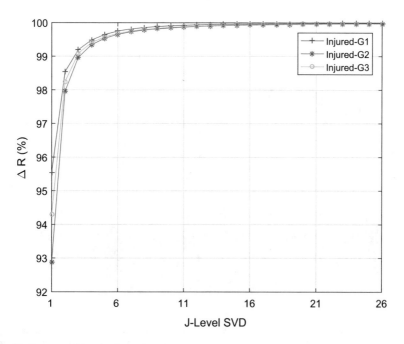

Fig. 4.7 Decomposition levels vs singular spectrum rate

and a mIA of 91.6%. The prediction of the same series via SWT-MIMO for 13-weeks ahead prediction is shown in Fig. 4.15; lower accuracy is observed with a $nRMSE$ of 10.1%, a $mNSE$ of 43.7% and a mIA of 71.8%.

The Injured-G2 prediction via MSVD-MIMO for 14-weeks ahead prediction is shown in Fig. 4.16a, b; from figures good fit is observed with a $nRMSE$ of 5.8%, a $mNSE$ of 81.9% and a mIA of 90.9%. The prediction of the same series via SWT-MIMO for 13-weeks ahead prediction is shown in Fig. 4.17, lower accuracy is observed with a $nRMSE$ of 19.6%, a $mNSE$ of 36.4% and a mIA of 68.2%.

The Injured-G3 prediction via MSVD-MIMO for 14-weeks ahead prediction is shown in Fig. 4.18a, b; from figures good fit is observed with a $nRMSE$ of 3.8%, a $mNSE$ of 81.4% and a mIA of 90.7%. The prediction of the same series via SWT-MIMO for 13-weeks ahead prediction is shown in Fig. 4.19, lower accuracy is observed with a $nRMSE$ of 13.0%, a $mNSE$ of 35.0% and a mIA of 67.5%.

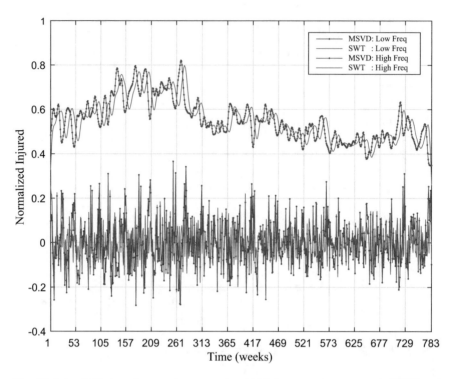

Fig. 4.8 Injured-G1, Low frequency component and high frequency component via MSVD and SWT

4.4   Empirical Application IV: MSVD and SWT for Multi-Step Ahead Fisheries Forecasting

4.4.1   Background

In the previous empirical application it was demonstrated that MSVD-AR model achieves the best accuracy for multi-step ahead forecasting of injured in traffic accidents in comparison with SWT-MIMO.

In this empirical application, the performance of the models based on MSVD and SWT is evaluated with the time series of fisheries which were predicted by means of HSVD and SSA in the previous chapter.

Only the MIMO strategy is now implemented due to it presents the same accuracy reached by Direct strategy. The structure of the linear Autoregressive model used for both MIMO and Direct conserve the same relationships among the predictors. However MIMO is developed in a single simulation with an unique model, its output is a predictions matrix X of $N_t \times \tau$ dimension, where τ is the maximum prediction horizon. On the other hand Direct strategy implements τ models. Therefore MIMO reduce the complexity of multi-horizon predictions in comparison with direct strategy.

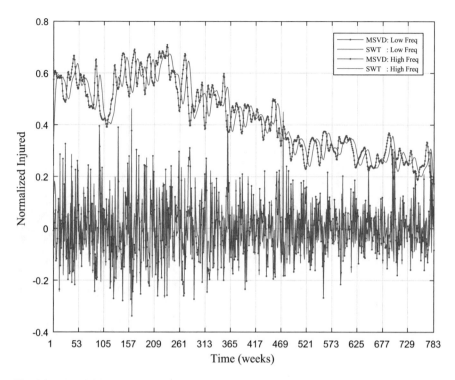

Fig. 4.9 Injured-G2, low frequency component and high frequency component via MSVD and SWT

4.4.2 Results and Discussion

4.4.2.1 Data Description

The stock of anchovy and sardine is in dependence of the environmental conditions; significant fluctuations of the marine ecosystem causes high stock variability. In this case study, the monthly nonstationary time series of anchovy and sardine catches are used, the data were collected by the National Service of fisheries and aquaculture (SERNAPESCA) [15], along the Chilean southern coast (33°S–36°S), from January 1958 for anchovy and from 1949 for sardine, to December 2011.

Considering the information of SERNAPESCA, it is observed that during the last decade the anchovy stock had been decreasing, while the sardine stock had been increasing; it is advisable to keep track of these species and their evolution, which is essential for exploitation policies and control rules.

Figures 4.20a and 4.21a show the observed time series, whereas Figs. 4.20b and 4.21b show the Fourier Power Spectrum (FPS) of anchovy and sardine respectively. High data variability is observed during the analyzed period; For example anchovy presents uptrend until year 2008 followed by downtrend until

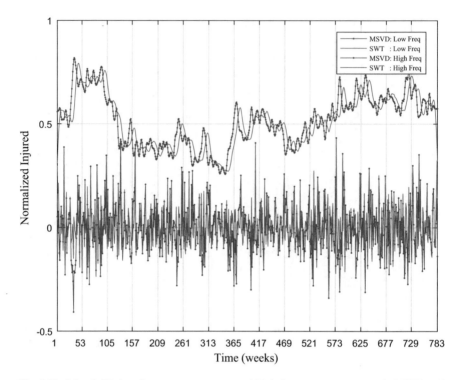

Fig. 4.10 Injured-G3, low frequency component and high frequency component via MSVD and SWT

end; whereas sardine shows scarcity from 1949 to 1989, followed by uptrend and abundance in the last years.

The FPS analysis shows the signal spectrum and the red noise spectrum, a signal spectrum peak is significative when its value is higher than the red noise spectrum [16]. Both series present the shortest significant peak of low frequency at month 30 at 65% of confidence level; besides 30 is a multiple harmonic of high frequency. The peaks information is used to set the AR model order ($P = 30$).

4.4.2.2   Decomposition in Intrinsic Components

The intrinsic components are extracted from the time series through the proposed MSVD method. The proposed methodology was illustrated in Fig. 4.2. The observed time series x is embedded in a Hankel matrix H through a fix window length $L = 2$, therefore H has $2 \times N - 1$ dimension. H is decomposed in singular values and singular vectors, and elemental matrices H_1 and H_2 are computed. Through unembedding the components of low frequency c_L and high frequency c_H are extracted.

Table 4.1 Multi-step MIMO forecasting results, $nRMSE$, $mNSE$, and mIA for Injured-G1

h (week)	nRMSE MSVD MIMO	SWT MIMO	mNSE (%) MSVD MIMO	SWT MIMO	mIA (%) MSVD MIMO	SWT MIMO
1	0.0021	0.34	99.9	98.0	99.9	99.0
2	0.0030	0.29	99.9	98.3	99.9	99.2
3	0.0007	0.44	99.9	97.6	99.9	98.8
4	0.0030	0.44	99.9	97.6	99.9	98.8
5	0.0003	0.56	99.9	96.8	99.9	98.4
6	0.0024	0.51	99.9	97.1	99.9	98.6
7	0.0041	0.63	99.9	96.5	99.9	98.3
8	0.0125	0.61	99.9	96.6	99.9	98.3
9	0.0366	5.6	99.8	67.8	99.9	83.9
10	0.0990	4.8	99.4	71.6	99.7	85.9
11	0.2512	6.5	98.5	64.2	99.3	82.1
12	0.6009	5.9	96.6	66.6	98.3	83.3
13	1.3685	10.13	92.2	43.7	96.1	71.8
14	2.9872	–	83.3	–	91.6	–
Min	0.0003	0.29	83.3	43.7	91.6	71.8
Max	2.9872	10.1	99.9	98.3	99.9	99.2
Mean 1–13 step	0.1834	2.8	98.9	84.1	98.9	92.0
Mean 1–14 step	0.3837	–	97.8	–	98.9	–

The process is iteratively performed until ΔR reaches its asymptotic point. Therefore each repetition gives new candidate components. The last repetition gives the definite components c_L and c_H. The Singular Spectrum Rate ΔR for each decomposition level J is illustrated in Fig. 4.22, the asymptotic value is reached when $\Delta R \approx 1$, and it was observed in the repetition 20 for anchovy and 18 for sardine.

SWT decomposition is implemented via Daubechies of order 2 (Db2) (due to the inaccurate results that were obtained with the other type of wavelet functions, they are not presented). Three decomposition levels ($J = 3$) were selected according with the period fluctuation between 8 and 16 months.

Figures 4.23 and 4.24, show the components of low frequency and high frequency obtained via MSVD and SWT for anchovy and sardine respectively. The c_L components extracted by both MSVD and SWT, show long-memory periodicity features, whereas the c_H components show short-term periodic fluctuations.

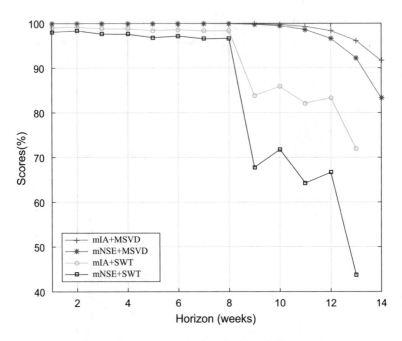

Fig. 4.11 Multi-step forecasting results, comparison for Injured-G1

Table 4.2 Multi-step MIMO forecasting results, $nRMSE$, $mNSE$, and mIA for Injured-G2

h (week)	RMSE		mNSE (%)		mIA (%)	
	MSVD MIMO	SWT MIMO	MSVD MIMO	SWT MIMO	MSVD MIMO	SWT MIMO
1	0.00048	0.62	99.9	97.9	99.9	98.9
2	0.00009	0.56	99.9	98.1	99.9	99.1
3	0.00132	0.85	99.9	97.2	99.9	98.6
4	0.00058	0.83	99.9	97.2	99.9	98.6
5	0.00105	1.12	99.9	96.3	99.9	98.2
6	0.00249	1.05	99.9	96.6	99.9	98.3
7	0.00761	1.22	99.9	95.9	99.9	97.9
8	0.02418	1.18	99.9	96.0	99.9	98.0
9	0.07071	10.4	99.8	65.6	99.9	82.8
10	0.19062	9.5	99.4	68.8	99.7	84.4
11	0.48291	12.5	98.5	58.6	99.2	79.3
12	1.15837	11.6	96.3	60.6	98.2	80.3
13	2.64161	19.6	91.6	36.4	95.8	68.2
14	5.77482	–	81.9	–	90.9	–
Min	0.00009	0.56	81.9	36.4	90.9	68.2
Max	5.77482	19.6	99.9	98.1	99.9	99.1
Mean 1–13 step	0.35	5.5	98.9	81.9	99.4	90.9
Mean 1–14 step	0.74	–	97.7	–	98.8	–

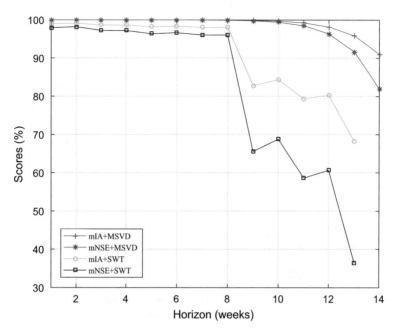

Fig. 4.12 Multi-step forecasting results, comparison for Injured-G2

Table 4.3 Multi-step MIMO forecasting results, $nRMSE$, $mNSE$, and mIA for Injured-G3

h (week)	nRMSE		mNSE (%)		mIA (%)	
	MSVD MIMO	SWT MIMO	MSVD MIMO	SWT MIMO	MSVD MIMO	SWT MIMO
1	0.0001	0.48	99.9	97.7	99.9	98.9
2	0.0020	0.40	99.9	98.0	99.9	99.0
3	0.0005	0.57	99.9	97.2	99.9	98.6
4	0.0007	0.57	99.9	97.3	99.9	98.6
5	0.0004	0.72	99.9	96.5	99.9	98.3
6	0.0015	0.63	99.9	96.9	99.9	98.4
7	0.0052	0.82	99.9	96.1	99.9	98.0
8	0.0165	0.82	99.9	96.1	99.9	98.1
9	0.0482	7.83	99.8	62.5	99.9	81.3
10	0.1303	6.59	99.4	67.7	99.7	83.9
11	0.3293	8.44	98.4	59.6	99.2	79.8
12	0.7864	7.96	96.2	61.8	98.1	80.9
13	1.7868	13.04	91.4	35.0	95.7	67.5
14	3.8944		81.4		90.7	
Min	0.0001	0.56	81.4	35.0	90.7	67.5
Max	3.89	19.6	99.9	98.0	99.9	99.0
Mean 1–13 step	0.24	5.46	98.9	81.7	99.4	90.9
Mean 1–14 step	0.5	–	97.6	–	98.8	–

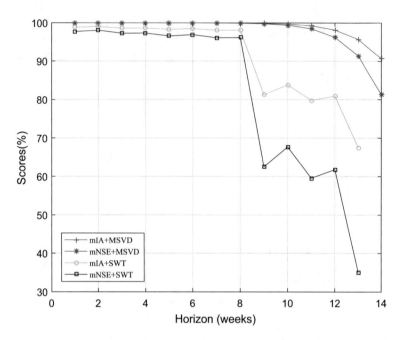

Fig. 4.13 Multi-step forecasting results, comparison for Injured-G3

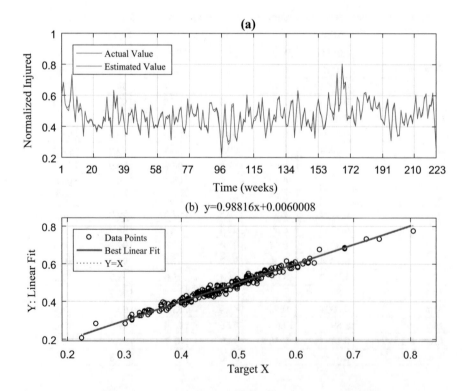

Fig. 4.14 Injured-G1 prediction by MSVD-MIMO (**a**) Observed vs predicted, (**b**) Linear fit

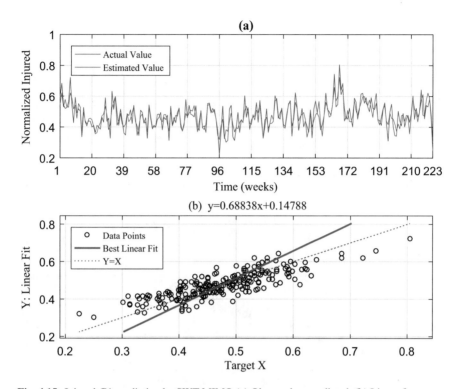

Fig. 4.15 Injured-G1 prediction by SWT-MIMO (**a**) Observed vs predicted, (**b**) Linear fit

4.4.2.3   Prediction Based on Intrinsic Components

The autoregressive models are implemented to predict the catches of anchovy and sardine for multiple horizon. The spectral analysis developed through FPS informed about the order of the model, it was shown in Figs. 4.20b and 4.21b. Each data set of low and high frequency has been divided into two subsets, training and testing. The training subset involves 80% of the samples, and consequently the testing subset involves the remaining 20%.

The inputs of both autoregressive models MSVD-MIMO and SWT-MIMO are the P lagged values of c_L and the P lagged values of c_H, and the outputs are the stock of anchovy or sardine for the next τ months.

The prediction performance is evaluated with efficiency metrics $nRMSE$, $mNSE$, and mIA. The results are presented in Tables 4.4 and 4.5 for anchovy and sardine respectively.

From Table 4.4 and Fig. 4.25, both models present good approximation. MSVD-MIMO model presents the best accuracy in comparison with SWT-MIMO. MSVD-MIMO obtains a significative gain in each forecasting horizon. The mean gain for 1–15 months of MSVD-MIMO over SWT-MIMO is 1.9% in $mNSE$ and 0.95% in mIA. The best gain of MSVD-MIMO over SWT-MIMO is for 12-months ahead

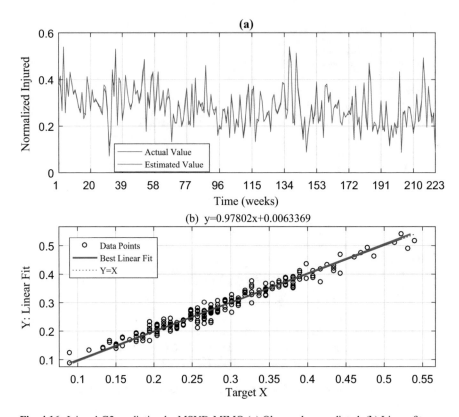

Fig. 4.16 Injured-G2 prediction by MSVD-MIMO (**a**) Observed vs predicted, (**b**) Linear fit

prediction, with 6.7% in $mNSE$ and 3.2% in mIA which is relevant for the anchovy annual fleet.

The best accuracy that was obtained previously for anchovy via MSVD-MIMO is now also obtained for Sardine. From Table 4.5 and Fig. 4.26, MSVD-MIMO achieves a significative gain in each forecasting horizon. The mean gain for 1–15 months of MSVD-MIMO over SWT-MIMO is 2.2% in $mNSE$ and 1.1% in mIA. The major gain of MSVD-MIMO over SWT-MIMO was reached for 12-months ahead prediction with 8.7% in $mNSE$ and 4.1% in mIA which is relevant for the sardine annual fleet.

The Anchovy prediction via MSVD-MIMO for 15-months ahead prediction is shown in Fig. 4.27. From figure good fit is observed between actual and estimated values. Metrics computation gives a $nRMSE$ of 23.9%, a $mNSE$ of 84.0% and a mIA of 92.0%. The prediction of the same series via SWT-MIMO for 15-months ahead prediction is shown in Fig. 4.28, lower accuracy is observed with a $nRMSE$ of 28.2%, a $mNSE$ of 80.9% and a mIA of 90.5%.

The Sardine prediction via MSVD-MIMO for 15-months ahead prediction is shown in Fig. 4.29. From figure good fit is observed between actual and estimated

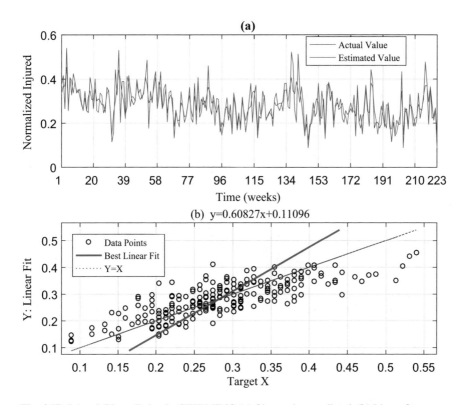

Fig. 4.17 Injured-G2 prediction by SWT-MIMO (**a**) Observed vs predicted, (**b**) Linear fit

values. From metrics computation, the prediction reaches a *RMSE* of 24.6%, a *mNSE* of 82.9% and a *mIA* of 91.4%. The prediction of the same series via SWT-MIMO for 15-months ahead prediction is shown in Fig. 4.30, lower accuracy is observed with a *RMSE* of 26.1%, a *mNSE* of 81.9%, and a *mIA* of 90.9%.

Conventional ANN-based forecast was also performed, the results are deficient for both anchovy and sardine. The best approximation via ANN was observed for 1-month ahead forecasting presents with a *MAPE* of 13.2% for anchovy and 16.5% for sardine; and a *mNSE* of 91.6% for anchovy and 85.2% for sardine. The regarding results for larger horizons were poor for both species.

4.5 Chapter Conclusion

In this chapter was presented the design of MSVD. Components of low and high frequency were obtained from nonstationary time series coming from transportation sector and fisheries sector.

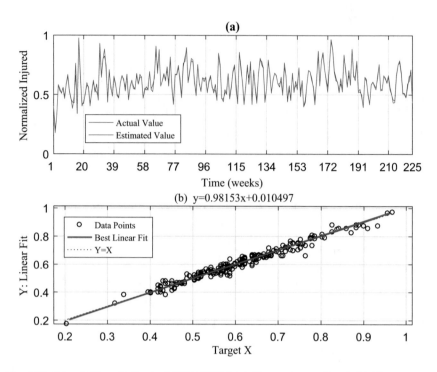

Fig. 4.18 Injured-G3 prediction by MSVD-MIMO (**a**) Observed vs predicted, (**b**) Linear regression, fitted line

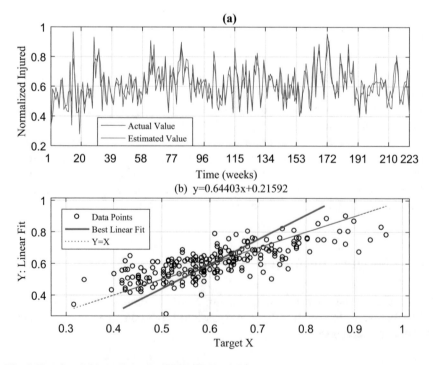

Fig. 4.19 Injured-G3 prediction by SWT-MIMO (**a**) Observed vs predicted, (**b**) Linear regression, fitted line

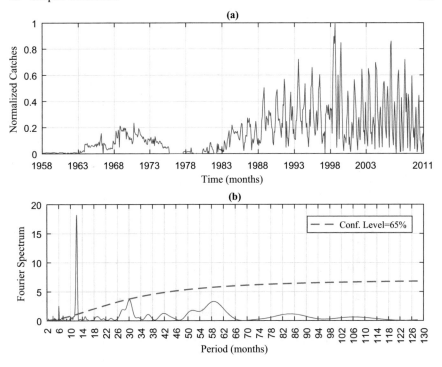

Fig. 4.20 (**a**) Anchovy catches, (**b**) FPS of Anchovy catches series

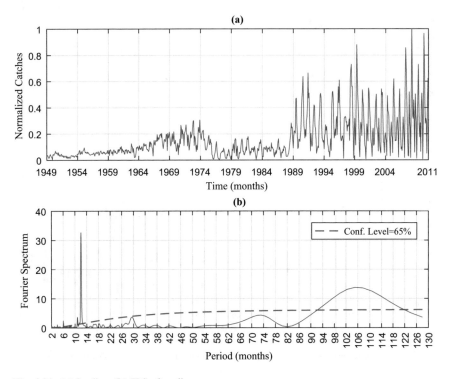

Fig. 4.21 (**a**) Sardine, (**b**) FPS of sardine

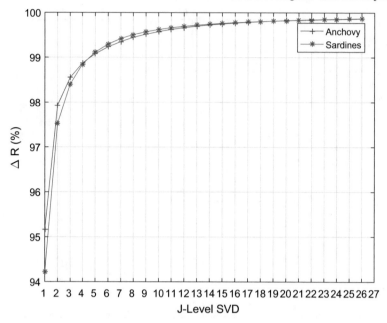

Fig. 4.22 Decomposition levels vs singular spectrum rate

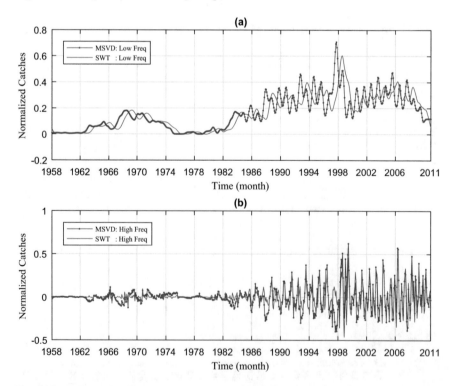

Fig. 4.23 Anchovy, low frequency component and high frequency component via MSVD and SWT

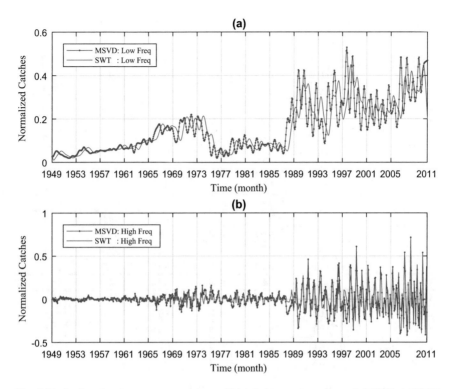

Fig. 4.24 Sardine, low frequency component and high frequency component via MSVD and SWT

The implementation of MSVD presents simplicity by the use of a fixed window length in the embedding step. Although the algorithm is iterative, the stop condition is guarantee by the convergence of ΔR parameter. MSVD requires less decisions with respect to SWT, which requires the selection of the wavelet function and the number of decomposition levels.

Both empirical application that were developed in this chapter probed the superiority of the proposed model MSVD-MIMO with respect to SWT-MIMO. Efficiency criteria $nRMSE$, $mNSE$ and mIA shown high accuracy of MSVD-based AR multi-step ahead forecasting, and superiority with respect to SWT-based AR multi-step ahead forecasting of both analyzed groups, injured persons and pelagic species catches.

MSVD-MIMO reaches an average $mNSE$-gain of 19.8% and an average mIA-gain of 8.9% for 13-weeks ahead forecasting of injured persons in traffic accidents. Whereas for anchovy and sardine predictions it was observed an average $mNSE$-gain of 7.7% and an average mIA-gain of 3.6% with respect to SWT-MIMO.

Table 4.4 Multi-step MIMO forecasting results, $nRMSE$, $mNSE$, and mIA for Anchovy

	nRMSE		mNSE (%)		mIA (%)	
h (month)	MSVD	SWT	MSVD	SWT	MSVD	SWT
1	0.002	0.12	99.9	99.9	99.9	99.9
2	0.003	0.26	99.9	99.8	99.9	99.9
3	0.011	0.59	99.9	99.6	99.9	99.8
4	0.003	0.94	99.9	99.4	99.9	99.7
5	0.005	0.99	99.9	99.4	99.9	99.7
6	0.020	0.99	99.9	99.3	99.9	99.7
7	0.056	1.73	99.9	98.7	99.9	99.3
8	0.187	2.78	99.9	98.1	99.9	99.0
9	0.538	2.87	99.6	97.9	99.8	98.9
10	1.357	3.71	99.1	97.3	99.5	98.7
11	2.674	9.51	98.0	93.2	99.0	96.6
12	4.887	13.36	96.3	90.2	98.1	95.1
13	8.119	15.04	93.9	89.5	96.9	94.7
14	14.25	14.89	89.8	89.4	94.9	94.7
15	23.99	28.17	84.0	80.9	92.0	90.5
Min	0.002	0.12	84.0	80.9	92.0	90.5
Max	23.99	28.17	99.9	99.9	99.9	99.9
Mean 1–12 step	0.81	3.15	99.4	97.7	99.7	98.9
Mean 1–15 step	3.74	6.39	97.4	95.5	98.7	97.8

Table 4.5 Multi-step MIMO forecasting results, $nRMSE$, $mNSE$, and mIA for Sardine

	nRMSE		mNSE (%)		mIA (%)	
h (month)	MSVD	SWT	MSVD	SWT	MSVD	SWT
1	0.001	0.099	99.9	99.9	99.9	99.9
2	0.002	0.268	99.9	99.8	99.9	99.9
3	0.004	0.727	99.9	99.5	99.9	99.8
4	0.001	1.156	99.9	99.2	99.9	99.6
5	0.004	1.102	99.9	99.3	99.9	99.6
6	0.018	1.185	99.9	99.3	99.9	99.6
7	0.057	2.209	99.9	98.6	99.9	99.3
8	0.144	3.706	99.9	97.5	99.9	98.8
9	0.330	3.687	99.8	97.5	99.9	98.7
10	0.764	4.476	99.4	96.9	99.7	98.5
11	1.745	8.024	98.7	93.9	99.4	96.9
12	3.195	13.542	97.4	89.6	98.7	94.8
13	7.337	13.879	94.9	89.9	97.4	94.9
14	14.809	16.498	90.3	89.1	95.2	94.6
15	24.605	26.109	82.9	81.9	91.4	90.9
Min	0.001	0.099	82.9	81.9	91.4	90.9
Max	24.6	26.1	99.9	99.9	99.9	99.9
Mean 1–12 step	0.52	3.35	99.6	97.6	99.8	98.8
Mean 1–15 step	3.53	6.44	97.5	95.5	98.8	97.7

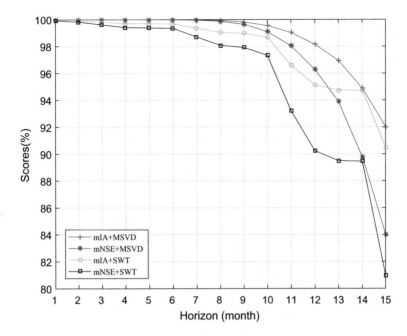

Fig. 4.25 Multi-step forecasting results, comparison for Anchovy

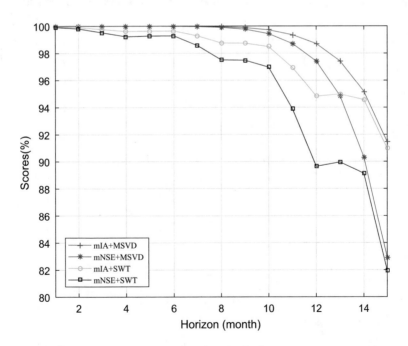

Fig. 4.26 Multi-step forecasting results, comparison for Sardine

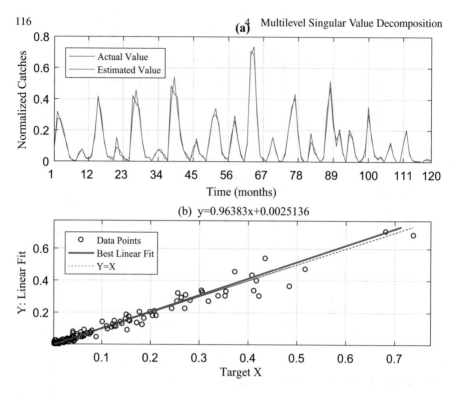

Fig. 4.27 Anchovy prediction by MSVD-MIMO (**a**) Observed vs predicted, (**b**) Linear fit

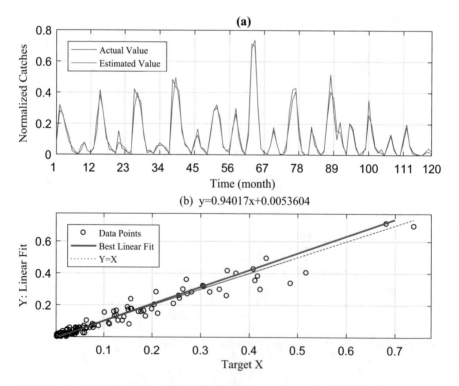

Fig. 4.28 Anchovy prediction by SWT-MIMO (**a**) Observed vs predicted, (**b**) Linear fit

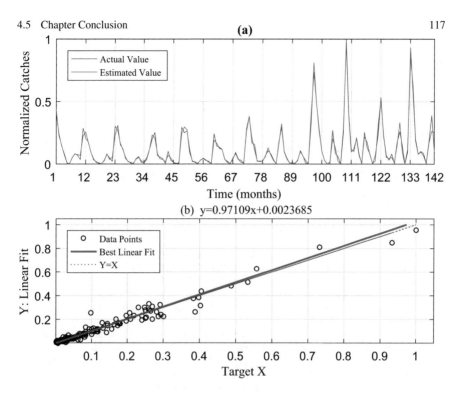

Fig. 4.29 Sardine prediction by MSVD-MIMO (**a**) Observed vs predicted, (**b**) Linear fit

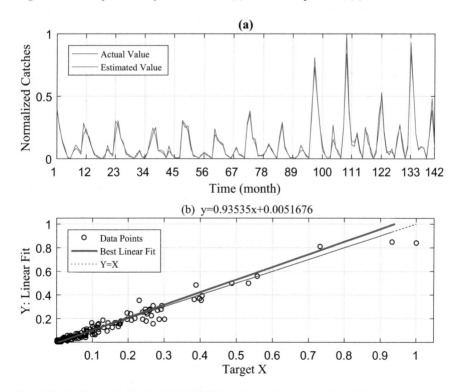

Fig. 4.30 Sardine prediction by SWT-MIMO (**a**) Observed vs predicted, (**b**) Linear fit

References

1. Mallat S (1989) A theory for multiresolution signal decomposition: the wavelet representation. IEEE Trans Pattern Anal Mach Intell 11(7):674–693
2. Barba L, Rodríguez N (2017) A novel multilevel-SVD method to improve multistep ahead forecasting in traffic accidents domain. Comput Intell Neurosci 2017:12 p, Article ID 7951395
3. Grossmann A, Morlet J (1984) Decomposition of Hardy functions into square integrable wavelets of constant shape. SIAM J Math 15:723
4. Magosso E, Ursino M, Zaniboni A, Gardella E (2009) A wavelet-based energetic approach for the analysis of biomedical signals: application to the electroencephalogram and electro-oculogram. Appl Math Comput 207(1):42–62
5. Haar A (1910) Zur theorie der orthogonalen funktionensysteme. Math Ann 69(3):331–371
6. Daubechies I (1988) Orthonormal bases of compactly supported wavelets. Commun Pure Appl Math 41(7):906–996
7. Shoaib M, Shamseldin AY, Melville BW (2014) Comparative study of different wavelet based neural network models for rainfall runoff modeling. J Hydrol 515:47–58
8. Daubechies I (1992) Ten lectures on wavelets CBMS - NSF regional conference series in applied mathematics. SIAM: Society for Industrial and Applied Mathematics, Philadelphia
9. Addison P (2002) The illustrated wavelet transform handbook. Institute of Physics Publishing, Bristol
10. Saâdaoui F, Rabbouch H (2014) A wavelet-based multiscale vector-ANN model to predict comovement of econophysical systems. Expert Syst Appl 41(13):6017–6028
11. Nason GP, Silverman BW, Georges O (1995) The stationary wavelet transform and some statistical applications. Wavelets and statistics. Springer, New York, pp 281–299
12. Shensa MJ (1992) The discrete wavelet transform: wedding the a trous and Mallat algorithms. IEEE Trans Signal Process 40(10):2464–2482
13. Coifman RR, Donoho DL (1995) Translation-invariant de-noising. Wavelets and statistics. Springer, New York, pp 125–150
14. Barba L, Rodríguez N (2016) Hybrid models based on singular values and autoregressive methods for multistep ahead forecasting of traffic accidents. Math Probl Eng 2016:14 p, Article ID 2030647
15. SERNAPESCA, Servicio Nacional de Pesca y Acuicultura. http://www.sernapesca.cl/. visited 12/2013
16. Torrence C, Compo GP (1998) A practical guide to wavelet analysis. Bull Am Meteorol Soc 79(1):61–78

Appendix A
Matlab Code: Chapter One

```
%=================================================%
%Code to generate and plot Gaussian White Noise
clear
close all
y1=wgn(1000,1,0);
figure
subplot(1,2,1)
plot(1:1000,y1)
xlabel('t')
ylabel('f(t)')
grid on
title('Gaussian White Noise')
subplot(1,2,2)
autocorr(y1)
title('Autocorrelation Function')
grid on
%=================================================%


%=================================================%
%Code to load an Ms Excel file, smooth the time series with moving average filter
and plot it.
clear
close all
range1='A1:A531';
y=xlsread('Data01.xls',1,range1); %People injured, Valparaiso time series
mabsy=max(abs(y));
y=y./mabsy;%Data normalization
ys=smooth(y,3);% smoothing by moving average filter using 3 points
t=1:numel(y);
```

© Springer International Publishing AG, part of Springer Nature 2018

L. M. Barba Maggi, *Multiscale Forecasting Models*,

https://doi.org/10.1007/978-3-319-94992-5

```
figure
plot(t,y,'-b',t,ys,'.-r')
xlabel('t (weeks)','FontName','times','FontSize',12)
ylabel('f(t)','FontName','times','FontSize',12)
legend('Oserved','Smoothed(MA-3)')
axis([1 531 0 1]);
grid on
%===============================================%


%===============================================%
%Code to identify the effective number of lags.
clear
close all
range1='A1:A531';
y=xlsread('Data01.xls',1,range1); %People injured, Valparaiso time series
mabsy=max(abs(y));
y=y./mabsy;
Perc=0.8; %80% data for training, 20% for testing
ys=smooth(y,3);% smoothing by moving average filter using 3 points
lags=40; %maximum number of lags
for p=1:lags
xd=lagdata(ys,p);% create the lagmatrix
[xe,ye,xv,yv]=Datatrntst(xd,y,p,Perc);%Function to split training sample and testing
sample
[yest,wl]=ARM(xe,ye,xv);% AR model, function to estimate the parameters with
LS method
Res1(p,:)=metricaT(yv,yest,p);%Function to compute the estimation accuracy
end
figure
plot(Res1(:,end),'.-b','LineWidth',0.5,'MarkerSize',10)
xlabel('Lag','FontName','times','FontSize',12)
ylabel('GCV','FontName','times','FontSize',12)
[a, b]=min(Res1(:,end))
grid on
%===============================================%
%===============================================%
%Code to crete the lagmatrix
function z=lagdata(x,Lag)
N=numel(x);
s=lagmatrix(x,[1:Lag]);
z=s(Lag+1:end,:);
end
%===============================================%
```

```
%================================================%
%Code to split the data in training and testing
function [xe,ye,xv,yv]=Datatrntst(X,Y,Lag,Perc)
Y=Y(Lag+1:end);
L=ceil(numel(Y)*Perc);
xe=X(1:L,:);
ye=Y(1:L);
xv=X(L+1:end,:);
yv=Y(L+1:end);
end
%================================================%
%================================================%
%Code to estimate the AR model coefficients
function [Y, w]=ARM(xe,ye,xv)
w=pinv(xe)*ye;
Y=xv*w;
end
%================================================%
%================================================%
%Code to evaluate the prediction accuracy.
function Rt=metricaT(yv,z,i)
er=yv-z;
Param=i;
Ns=numel(yv);
MSE1=mean(er.*er);
RMSE=sqrt(MSE1);
PRMSE=RMSE*100;
R2=(1-(var(er)/var(yv)))*100;
MAE=mean(abs(er));
MAPE=mean(abs(er./yv))*100;
GCV=RMSE/(1 - (Param/Ns)).^2;
Rt=[i MAE MAPE PRMSE R2 GCV];
end
%================================================%

%================================================%
%Code to identify the effective number of hidden nodes
function eNh=HiddenSet(xe,ye,xv,yv,Epoch,p)
for Nh=1:20
[net1]=p1lm(Nh,xe',ye',Epoch);
netsNh=net1;
Xest=sim(net1,xv');
Xest=Xest';
Res1(Nh,:)=metricaT(yv,Xest,Nh,p)
end
```

```
[a, b] = min(Res1(:, end));
eNh=b;
end
%===============================================%


%===============================================%
%Code to implement Resilient Backpropagation Algorithm
function [net]=p1rprop(Nh,P,T,Epoch)
net=newff(P,T,[Nh 1],'logsig','purelin','trainrp');
net.trainParam.epochs=Epoch;%Maximum number of epochs to train
net.trainParam.show=25;%Epochs between displays (NaN for no displays)
net.trainParam.goal=1e-5;%Performance goal
net.trainParam.showCommandLine=0;%Show Command Line
net.trainParam.showWindow=0;%Show training GUI
net.trainParam.max_fail=5;%Maximum validation failures
net.trainParam.min_grad=1e-10;%Minimum performance gradient
net.trainParam.time=inf;%Maximum time to train in seconds
net.trainParam.lr=0.01;%Learning rate
net.trainParam.delt_inc=1.2;%Increment to weight change
net.trainParam.delt_dec=0.5;%Decrement to weight change
net.trainParam.delta0=0.07;%Initial weight change
net.trainParam.deltamax=50.0;%Maximum weight change
net.performfcn='msereg';%Performance function
[net]=train(net,P,T);%ANN configuration
%===============================================%


%===============================================%
%Code to implement Levenberg-Marquardt Algorithm
function [net]=p1lm(Nh,P,T,Epoch)
net=newff(P, T, [Nh 1],'logsig','purelin','trainlm');
net.trainParam.epochs=Epoch;%Maximum number of epochs to train
net.trainParam.show=50;%Epochs between displays (NaN for no displays)
net.trainParam.goal=0;%Performance goal
net.trainParam.showWindow=0;%Show training GUI
net.trainParam.max_fail=5;%Maximum validation failures
net.trainParam.min_grad=1e-10;%Minimum performance gradient
net.trainParam.mu=0.001;%Initial mu
net.trainParam.mu_dec=0.1;%mu decrease factor
net.trainParam.mu_inc=10;%mu increase factor
net.trainParam.mu_max=1e10;%Maximum mu
net.trainParam.time=inf;%Maximum time to train in seconds
net.performfcn='msereg';%Performance function
[net]=train(net,P,T);%ANN configuration
%===============================================%
```

Appendix B
Matlab Code: Chapter Two

```
%=================================================%
%Code to implement HSVD
clear
close all
y=xlsread('Accidentes.xls',1,'b1:b783');
y=y./max(abs(y));
N=length(y);
k=7;%number of rows of matrix H (effective window length
%Embedding and SVD
[C]=hsvd(k,y'); %C is the matrix of components
CL=C(1,:)'; %component of low frequency
CH=y-CL; %component of high frequency
t=1:N;
figure
subplot(2,1,1)
plot(t,y,'-b',t,CL,'.-k')
xlabel('t (weeks)','FontName','times','FontSize',12)
ylabel('f(t)','FontName','times','FontSize',12)
legend('Oserved','C_L Component')
grid on
hold on

   subplot(2,1,2)
plot(t,CH,'.-r')
xlabel('t (weeks)','FontName','times','FontSize',12)
ylabel('f(t)','FontName','times','FontSize',12)
legend('C_H Component')
grid on
hold off
%=================================================%
```

© Springer International Publishing AG, part of Springer Nature 2018
L. M. Barba Maggi, *Multiscale Forecasting Models*,
https://doi.org/10.1007/978-3-319-94992-5

```
%=====================================================%
function [C]=hsvd(k,y)
% k number of rows of H
% y vector to decompose
N=numel(y);
L=N-k+1; %number of rows of H
for i=1:k
HA(i,:)=y(i:L+i-1);
end
for i=1:k
[U, S, V]=svd(HA,0);
s=diag(S);
vi=V(:,i)';
Ai=s(i)*U(:,i)*vi;
end;
C=zeros(k,N); %matrix of components
for i=1:k
c_1=Ai;
p1=numel(c_1(1,:));
C(i,1:p1)=c_1(1,:);
for j=2:k
temp2=c_1(j,end);
p1=p1+1;
C(i,p1)=[temp2];
end
end
end
```